# Owls - Clever Survivors

*Edited by Heimo Mikkola*

Published in London, United Kingdom

Owls - Clever Survivors
http://dx.doi.org/10.5772/intechopen.95727
Edited by Heimo Mikkola

Contributors
Samuel Pacenovsky, Alexander Kürthy, Alan Sieradzki, Heimo Juhani Mikkola, Anita Mikkola, Zlatozar Boev, Omar F. Al-Sheikhly, Ahmad E. Essam Aidek, Héctor Cadena-Ortiz, Jorge Brito, María Cristina Ríos, Paolo Piedrahita, Glenda Pozo-Zamora, Juan Freile, Hermann Wagner



First published in London, United Kingdom, 2023 by IntechOpen
IntechOpen is the global imprint of INTECHOPEN LIMITED, registered in England and Wales, registration number: 11086078, 5 Princes Gate Court, London, SW7 2QJ, United Kingdom

British Library Cataloguing-in-Publication Data
A catalogue record for this book is available from the British Library

Additional hard and PDF copies can be obtained from orders@intechopen.com

Owls - Clever Survivors
Edited by Heimo Mikkola
p. cm.
Print ISBN 978-1-80355-390-0
Online ISBN 978-1-80355-391-7
eBook (PDF) ISBN 978-1-80355-392-4

# Meet the editor

Heimo Mikkola is a bird watcher who has concentrated on owls since 1965. He did his master's study on the great grey owl and obtained a Ph.D. in the Ecological Relationships of European Owls. In 1988, he became an adjunct professor at the University of Eastern Finland. His life-long international career with the Food and Agriculture Organization (FAO) of the United Nations (FAO) began in Colombia in 1974. He retired in 2007 as the resident representative of FAO in Uruguay. Since retiring from FAO, Dr. Mikkola has been a part-time professor at three Kazakh universities and one Kyrgyz university. He has visited 139 countries and written almost 700 work reports and scientific papers, including ten edited books. In 2014, he was nominated as "Champion of Owls" in Houston, USA.

# Contents

**Preface** **IX**

**Section 1**
What Makes an Owl? **1**

**Chapter 1** **3**
Designed for Darkness: The Unique Physiology and Anatomy of Owls
*by Alan Sieradzki*

**Section 2**
Food Studies **27**

**Chapter 2** **29**
What Do We Know about the Diet of Ecuadorian Owls?
*by Héctor Cadena-Ortiz, Jorge Brito, María Cristina Ríos, Paolo Piedrahita, Glenda Pozo-Zamora, Hermann Wagner and Juan Freile*

**Section 3**
Voice Studies **49**

**Chapter 3** **51**
The Vocal Activity of Twelve African Owl Species
*by Heimo Mikkola and Anita Mikkola*

**Section 4**
Country Biodiversity, Population Changes and Conservation Studies **69**

**Chapter 4** **71**
Qualitative and Quantitative Changes in a Guild of Forest Owls: Eurasian Pygmy Owl (*Glaucidium passerinum*), Ural Owl (*Strix uralensis*), Tawny Owl (*Strix aluco*), Boreal Owl (*Aegolius funereus*) at Kamenný Hrb – Bankov Site in Volovské Mountains Near Košice Town, Eastern Slovakia, between Years 1989 and 2021
*by Samuel Pačenovský and Alexander Kürthy*

**Chapter 5** 89
Owls (Strigiformes Wagler, 1830) in Bulgaria: Past and Present
(A Review of the Fossil Record and Present Status of Recorded Species)
*by Zlatozar Boev*

**Chapter 6** 103
Queens of the Night, the Owls of Iraq and Syria - Species, Current
Distribution, and Conservation Status
*by Omar F. Al-Sheikhly and Ahmad E. Aidek*

**Section 5**
Eating Owls as Medicine or for Magic and Witchcraft 139

**Chapter 7** 141
Owls Used as Food and Medicine and for Witchcraft in Africa
*by Heimo Mikkola*

# Preface

Owl studies have often taken me to the best bat biotopes as well, and quite some time ago I started to collect data on bats and owls in global folklore. Bats and owls are both iconic nocturnal creatures that are surrounded by a myriad of strange old wives' tales and superstitions worldwide. This volume is a collection of global perceptions aiming to promote a better biocultural richness for humans, bats, and owls as our long-term nocturnal companions.

In ancient Azania in north and central Africa bats were universally disliked and considered very unlucky if found hooting around a homestead during the night [1]. Also, in central Africa should a bat or an owl come near the house, or a bush cat defecate in the compound, the owner must go at once to a diviner to discover what remedies must be taken to ward off the evil. A witch shape is believed to be capable of sucking the life out of a sleeping man or woman [2]. In Nilotic Sudan, witchcraft was usually performed at night and therefore owls and bats were associated with it [3].

In southeastern Australia, tribal life was much bound up with animals, but men were especially represented by bats and women by owls. The Wotjobaluk tribe held the firm belief that the lives of its women were influenced by the owl Yàrtatgurk, and those of men by the bat: "the common bat belongs to the men who protect it against injury, even to half-killing their wives for its sake" [4].

The fern owl belongs to the women and although the bird of evil omen, creates terror at night by its cry, it is jealously protected by them. "If a man kills one owl even by mistake, they are enraged as if it was one of their children and women will strike him with their yam sticks." The jealous protection thus afforded by Australian men and women to bats and owls respectively is not based purely on selfish considerations. Each woman believes that the lives of her mother, sister, daughter, and so forth, equally her own, are bound up with the lives of particular owls and that in guarding the owl species, she is guarding the lives of her female relations. Females say that "if my sister Mary's life is in an owl, the owl is my sister and Mary is an owl." Respectively, males said, "if my brother John's life is in a bat, then, on the other hand, the bat is my brother as well as John." Since no one knew exactly which bat or owl guarded a particular soul, all bats and owls were effectively protected [4].

In parts of the Indian sub-continent people believed that Bat was married to Owl [5].

In eastern Indonesia, Nage people believe that all witch birds like owls and even diurnal raptors are flesh eaters, either hunters or scavengers, while bats are not carnivorous. They also describe insect-eating small bats as fruit-consuming animals. The Nage do not taboo the consumption of bat flesh like they do with flesh of all witch birds. They categorize bats as "birds," although they don't consider them as good examples of birds. Bats are not identified with any category of spiritual beings, nor are they prominent in myth. More importantly, bats are not included in the symbolic class of witch birds. The Nage ascription to witches of nocturnal habits, killing and cannibalistic consumption, eerie vocalizations, flight, and the ability to rotate their heads would appear, in most if not quite all respects, to reflect the empirically observable physical features of owls rather than of human beings [6].

It was striking to learn that in the old European folklore owls and bats were often mixed, causing many strange beliefs [7]. First it was believed that bat is a bird, and to add confusion it was named as strix in the folktales. The strix was regarded as a veritable bird of hell, and Gaius Plinius Secundus (AD 23/24–79) wrote that the strix was potent also in malediction [7].

In Ancient Greek story of Polyphonte who gave birth to two humanoid bear-like sons because of her union with a bear. The bear twins honored neither men nor gods; rather, they were cannibals who attacked strangers on the road. Zeus despised the sons and their mother and asked Hermes to punish them [8]. Mother Polyphonte was transformed into a strix that cries by night, without food or drink, with head below and tips of feet above, a harbinger of war and civil strife to men. Son Oreius became a bird that is seen for no good and the other son Agrius was changed into a vulture, of all birds most detested by gods and men and possessed of a constant craving for human flesh and blood. Antonius cites Boeus's second book, *The Origin of Birds*, as the source of this story from the end of the fourth century BC; however, Boeus' work has been lost [7]. To the ancients in general, the bat is a bird. Therefore, it has been questioned if the strix was a bat and not an owl as so often supposed [8].

The ancient literature of Greece and Rome during the seven centuries from Boio to Serenus Sammonicus (died 212 AD) present a consistent view of the strix. The bird is clearly mythical, but its physical characteristics were those of a bat rather than those of an owl, as so often supposed. The bird is vampiric, but never a true vampire. Titinius (lived around 170 BC) makes it clear that strix is plainly chiropterous. Domenico Comparetti (1835-1927) also wrote in his 'Novelline Populari Italiene' that the souls of the three beautiful sisters were three pipistrelle bats [7].

Later when Striges became the ornithological appellation of the entire suborder of the owls, the confusion was even deeper. Públius Ovidius Nāsō (43 BC-17/18 AD) mentions eggs of the strix and Quintus Horatius Flaccus (65 BC – 8 BC) and Sextus Propertius (50-45 BC – ca. 15 BC) mention its feathers. Literally, eggs and feathers exclude the mammalian, chiropteran strix of Titinius, but it was natural to ascribe them to any bird, real or imaginary. Irish folk-lore still believes in the eggs of bats and their potency in malignant charms. Pomponius Porphyrio (flourished 2nd century AD) sees strix as a vague and undefined "nocturnal bird of evil omen" but there is no compelling evidence for the screech owl [7].

In Medieval Icelandic and Viking Age Norse mythology the bat is the messenger of Hel, the goddess of darkness and death, and is feared as such [7, 9]. The owl relations of Vikings is not so well documented, but it is said that the Vikings and their ancestors looked at animals with awe and administration [10]. Huathe, the hawthorn, is a mystic rune associated with the owl. In the Viking world, the owl is the guide to the underworld. It helps people to see the spiritual and actual dark and it shows how to look inside the darkness in ourselves and find a way out of it [11].

The earliest inhabitants of Jamaica, the Taino people, believed that mankind originated from caves. The bat and the owl were very important symbols in Taino mythology and death. The bat represented the opias because fruit-eating bats like Jamaican fruit-eating bat *Artibeus jamaicensis* love eating guavas. Guava is also the favorite food of the Taino spirits of the dead, explaining why bats are perceived as death images in the folklore [12]. Amongst Jamaican folk tradition, the owl also symbolizes death. The owl is considered the divine bird of the coyaba, heaven or underworld. Taino people were terrified of the owl's nocturnal call because they believed the bird was the herald of the lord of coyaba and it was delivering the message that a human life was about to end [13].

In South America, bats and owls have had agricultural as well as death associations for pre-Colombian people. In a Mochica vessel, a bat is posed as if it were presenting food. Mochica squash depictions may also have an owl head, another indication of the interaction of bat and owl, for vegetables seem to have distinct associations – a fanged deity with maize, a diseased face with potatoes, and so on [14].

The north coast of Peru is one of the regions where bat iconography is particularly prominent. On Mochica pottery an anthropomorphic bat is an agent of human sacrifice having funeral connotations. An anthropomorphic owl also appears in this role. Bats and owls are nocturnal, and both can be predators. In folklore from the South American lowlands at both ends of the continent, they are often brothers-in-law [15–20].

On the south coast of Peru there are fewer bats, and they are not presented in Nasca ceramics where the vencejo or the hummingbird may play a similar role [14]. The common potoo (*Nyctibius griseus*) would be an interesting substitution for the bat because it belongs to the order of Caprimulgiformes and greatly resembles an owl.

This book includes seven chapters that discuss owls and their significance in Africa, Bulgaria, Ecuador, Iraq, Slovakia, and Syria.

My warmest thanks to Author Service Manager Zrinka Tomicic at IntechOpen for her professional and helpful cooperation throughout the publication of this book. I am also grateful to Alan Sieradzki for his everlasting enthusiasm in seeking and finding some little-known bat and owl lore publications.

**Heimo Mikkola**
Eastern Finland University,
Kuopio, Finland

## References


[1] Evans-Pritchard EE. Witchcraft, Oracles and Magic among the Azande. Oxford: Clarendon Press; 1937. p. 265

[2] Jeffreys MDW. African pterodactyls. Journal of the Royal African Society. 1944;**43**(171):72-74

[3] Seligman CG, Seligman BZ. The Pagan Tribes of the Nilotic Sudan. London: Routledge G & Sons; 1932. p. 732

[4] Frazer JG. The Golden Bough. A Study in Magic and Religion. New York: MacMillan; 1922. p. 625

[5] Lewis D. World Owl Mythology—The Owl Pages. 2012 [Last updated: 08 October 2012]

[6] Forth G. Symbolic birds and ironic bats: Varieties of classification in Nage Folk Ornithology. Ethnology. 2009;**48**(2):139-159

[7] Oliphant SG. The story of Strix ancient. Transactions and Proceedings of the American Philological Association. 1913;**44**:133-149

[8] Wikipedia. Polyphonte. Available from: https://en.wikipedia.org/wiki/Polyphonte [Accessed: 10 December 2021]

[9] Wikipedia. Edda. Available from: https://en.wikipedia.org/wiki/Edda [Accessed: 10 December 2021]

[10] Dæhlen M. Vikings had a Completely Different Relationship to Animals than We have Today. Available from: Sciencenorway.no/animal-world-animals-history/Vikings-had-a-completely-different-r... [Accessed: 07 July 2021]

[11] Neenes. What Does the Owl Represent in Norse Mythology? Available from: https://neenes.com/what-does-the-owl-represent-in-norse-mythology [Accessed: 10 December 2021]

[12] Atkinson L-G. Taino Influence on Jamaican Folk Traditions. Taino Day Presentation in May. 2010. Available from: http://www.jnht.com/download/influence.pdf [Accessed: 07 August 2021]

[13] Arrom Gonzales JJ. Mitologia y artes Prehispánicas de las Antillas. 2nd ed. México: Siglo XXI Editores; 1989. p. 191. ISBN: 9682315522

[14] Benson EP. Bats in South American iconography. Andean Past. 1987;**1**:165-190

[15] Roth WE. An inquiry into the animism and Folk-lore of the Guiana Indians. In: Annual Report of the Bureau of American Ethnology 1908-1909. Vol. 30. 1915. pp. 103-386

[16] Lothrop SK. The Indians of Tierra del Fuego. Contributions from the Museum of the American Indian X. 1928

[17] Wilbert J. Folk Literature of the Warao Indians: Narrative Material and Motif Content. Los Angeles: Latin American Studies, University of California; 1970. p. 15

[18] Wilbert J. Yupa Folktales. Los Angeles: UCLA Latin American Center Publications; 1974

[19] Wilbert J. Folk Literature of the Selknam Indians. Los Angeles: UCLA Latin American Center Publications; 1975

[20] Wilbert J, Simoneau, K. Folk Literature of the Mataco Indians. Los Angeles: UCLA Latin American Center Publications; 1982

Section 1

# What Makes an Owl?

Chapter 1

# Designed for Darkness: The Unique Physiology and Anatomy of Owls

*Alan Sieradzki*

## Abstract

Owls are the only truly nocturnal avian raptors and have evolved several unique adaptations to perfectly fill this role. For example, their unique large tubular eyes, packed with light-sensitive cells, enable owls to operate in almost total darkness, while their remarkable auditory system allows them to operate in complete darkness. This unique and complex auditory system is a combination of specialised feathers forming a parabolic facial disc, adjustable operculum, or flaps and, in some species, asymmetrical ear openings. This unique system allows the owl's brain to construct an *auditory map of space* when locating its prey. As remarkable as it seems, this is tantamount to owls being able to hear in 3D. While there are minor variations in the individual physiology between certain species due to the type of prey they take and the habitat they operate in, all owls are subject to the same unique adaptations in eyes, hearing, head rotation, feather structure, digestive system, and hind limb musculature. In this study, we examine each individual adaptation that combines to make the owl a superbly designed nocturnal predator and also look at some shared mechanisms and behaviour patterns that are crucial to its survival.

**Keywords:** owl, adaptation, vision, hearing, digestion, anatomy, feathers, musculature, variation

## 1. Introduction

Owls are one of the most distinctive-looking birds in the world. With their upright stance, large head with forward-facing round eyes, flat facial disc and soft fluffy plumage, they cannot be mistaken for anything else. This distinctive outward appearance is the result of many unique evolutionary adaptations, which have enabled the owl to become a highly efficient crepuscular and nocturnal predator. While they share the night skies with insectivorous Caprimulgiformes, such as Nightjars, Frogmouths, Potoos and Oilbirds, and with Owlet-nightjars of Aegotheliformes, the true nocturnal owls are unlike their diurnal raptor counterparts. Whereas the diurnal raptors, consisting of Eagles, Falcons, Hawks, and Buzzards, have separately evolved in response to a wide range of prey species and habitats, the owl is singularly the only true nocturnal raptor. Species of owls are found on every continent and nearly in every country of the world, except Antarctica and some small isolated islands, and can thrive in habitats as diverse as frozen tundra, equatorial rainforests, temperate northern forests, and even open grasslands and deserts.

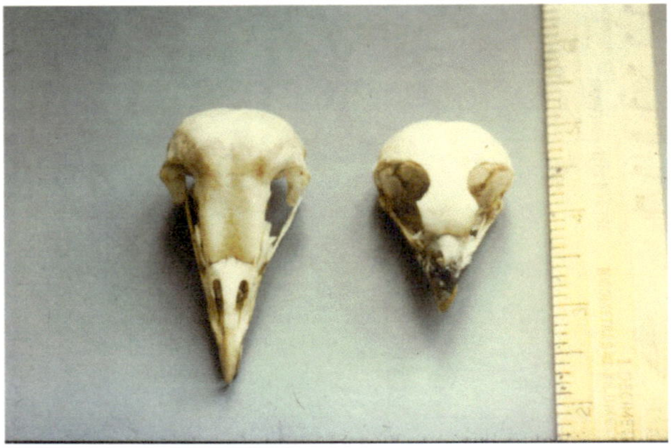

**Figure 1.**
*Tytonidae—Strigidae skull comparison. L—Barn owl (Tyto alba). R—Little owl (*Athene noctua*). Photo: Alan Sieradzki.*

Owls belong to the taxonomic order of Strigiformes, which is divided into the two families of Strigidae (Typical Owls) and the much smaller family of Tytonidae (Barn, Bay and Grass Owls). While there are some distinct anatomical differences between the two families, most notably in the structure of the skull (**Figure 1**), both families share the same adaptations that make them owls. There are approximately 250 known species of owl in the world, ranging in size from the diminutive Elf Owl (*Micrathene whitneyi*) to the enormous Eurasian Eagle Owl (*Bubo bubo*). While there are minor variations in the individual physiology between certain species due to the type of prey they take and the habitat they operate in, they are all subject to the same unique adaptations in eyes, hearing, head rotation, feathers structure, digestive system, and hind limb musculature.

Recent research has used a genome-wide scan to uncover the genetic and selective mechanisms that are the basis of the owl's unique sensory adaptations. As predicted, a primary finding of the study was that genes involved in sensory perception showed a genome-wide signal of positive selection. This category included genes involved in acoustic and light perception, photosensitivity, phototransduction, dim-light vision, and the development of the retina and inner ear. Genes involved in circadian rhythms, which regulate the body's internal clock, also showed evidence of accelerated evolution, as did some genes related to feather production [1].

## 2. Eyes

Arguably the most distinctive feature of owls is their large forward-facing eyes. Instead of the usual 'disc'-shaped eyes normally found in birds, owls have developed large 'tubular'-shaped eyes (**Figure 2**) that are held tight within the orbit and protected by the *scleral ossicles*, or 'sclerotic ring', composed of a series of small interlocking bones that form a bony ring within the sclera. These 'tubular' eyes are so large that in some small species of owl they can take up to 50% of the skull area. Some eagle species have developed 'tubular' eyes, but these are much shorter in length than an owl's, while the majority of diurnal raptors have much smaller 'globose' eyes (**Figure 3**) [2].

The owl's large, forward-facing eyes allow for considerable binocular vision, giving an excellent, but a fairly narrow field of view of 110 degrees, with an overlap of

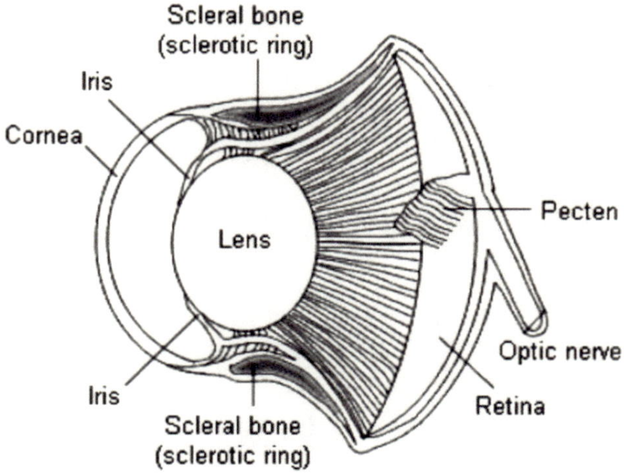


**Figure 2.**
*Diagram of an owl's eye. Image credit: owlpages.com*

approximately 70 degrees (man, by comparison, has a field of view of 180 degrees with an overlap of 140 degrees) [3]. With such a narrow field of view, many species resort to that very distinctive behaviour of owls, head bobbing, to accurately judge distance and position.

The eyes have extremely large cornea (the transparent outer coating of the eye) and pupil (the opening at the centre of the eye). A larger cornea allows for a larger pupil size, which in turn serves to increase the number of photons that reach the retina (light-sensitive tissue on which the image is formed), thereby improving visual sensitivity [4]. The pupil's size is controlled by the iris (the coloured membrane suspended between the cornea and lens). When the pupil is larger, more light passes through the lens and onto the large retina. Light sensitive retinal cells act as receptors and form images. These receptors are made up of two types of cells, rods and cones; so named for their shapes. Cones distinguish colours, function in bright light, and are needed for sharp resolution, while rods function in low light or at night and

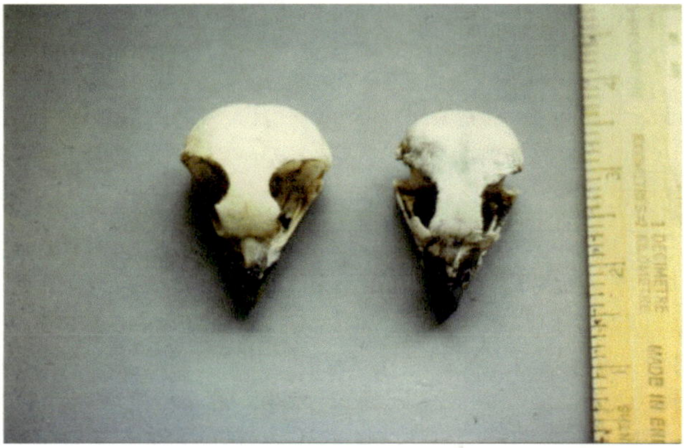

**Figure 3.**
*Eye orbit comparison nocturnal-diurnal. L—Little owl (*Athene noctua*). R—Kestrel (Falco tinnuculus). Photo: Alan Sieradzki.*

**Figure 4.**
*The nictitating membrane of the Eurasian eagle owl (*Bubo bubo*). Photo: Bruce Marcot.*

are sensitive to movement. Primarily a nocturnal predator, an owl's eyes are packed with rods, giving owls excellent nocturnal vision without the need of the *tapetum*, a reflecting layer at the back of the eye found in most nocturnal animals, including those other nocturnal birds, the Caprimulgiformes. While the ambient light on a cloudy, moonless night rarely drops below an illumination level of 0.004-foot candles, experiments have shown that some species, such as Long-eared Owl (*Asio otus*), Tawny Owl (*Strix aluco*), and Barn Owl (*Tyto alba*), can spot and approach prey from a distance of 6 feet or more under illumination as low as 0.000,00073 foot candles [5]. A recent study's findings indicate that owls may have independently evolved a DNA packaging mechanism in the retina that enhances light channelling in photoreceptors, a feature that has not been observed in any other bird species to date [1].

But an owl's eyes also contain enough cones to enable it to see perfectly in daylight; owls are by no means blind in the daylight. In fact, with its wide range of pupil adjustment, an owl's ability to see sharply is as developed as in any diurnal raptor and has allowed many owl species, such as the Burrowing Owl (*Athene cunicularia*), the Snowy Owl (*Bubo scandiacus*) and the Short-eared Owl (*Asio flammeus*), to become highly successful diurnal hunters. To protect their eyes, owls are equipped with three eyelids. They have a normal upper and lower eyelid, the upper closing when the owl blinks (as in humans—the only bird to do so), and the lower closing up for sleep. The third eyelid is called a *nictitating membrane*, which is a thin layer of tissue that closes diagonally across the eye, from the inside to the outside and cleans and protects the surface of the eye (**Figure 4**).

## 3. Head rotation

Because of the large size of the 'tubular' eyes and the fact that they are locked into place by a sclerotic ring of bone, ocular mobility in owls is virtually non-existent [6]. To compensate for this lack of eye movement, and a fairly narrow field of view, owls have evolved with the ability to laterally swivel the head smoothly and quickly through 270 degrees and vertically 90 degrees. Owls have 14 cervical vertebrae, but so do many other species of bird. Indeed, 14 is about the average number of cervical vertebrae in birds in general (birds can have between 10 and 26 vertical vertebrae depending on species) [7]. All birds have to have the ability to turn their heads through 180 degrees and more for preening. The secret to the owl's ability to swivel its head smoothly and quickly through 270 degrees in the manner that it does is down to two areas of adaptation. The first adaptation is to the neck itself. Owls have only one occipital articulation with the cervical vertebrae, while the neck is permanently compressed into a loose 'S' shape [8]. As with a spring coil, this gives the neck great flexibility. It has also been discovered that there are varying degrees of axial rotation within the individual intervertebral joints [9] and that the combination of yawing and rolling in sections of the cervical spine maximises head rotation [10]. The second adaptation is in the reinforcement of the walls of the oesophagus, trachea, and arteries to withstand the enormous torque involved as the head is turned through so many degrees. Also, it has recently been discovered that in the owl neck, one of the major arteries feeding the brain passes through bony holes in the vertebrae. These hollow cavities are approximately 10 times larger in diameter than the vertebral artery travelling through it. The extra space in the transverse foramina, as the holes surrounding the vertebral arteries, are known, creates a set of cushioning air pockets that allow the artery to move around when twisted. Twelve of the 14 cervical vertebrae in the owl's neck were found to have this adaptation. Blood vessels at the base of the head, just under the jaw bone, can also act as contractile blood reservoirs, allowing owls to pool blood to meet the energy needs of their large brains and eyes, while they rotate their heads. The supporting vascular network, with its many interconnections and adaptations, helps to minimise any interruption in blood flow [11].

## 4. Hearing

Owls have a unique, complex and highly developed, and specialised auditory system designed to aid in the location and capture of prey. Most owls use a combination of their remarkable hearing and eyesight to locate and capture their prey. However, some species, such as the Barn Owl (*Tyto alba*), the Great Grey Owl (*Strix nebulosa*), the Long-eared Owl (*Asio otus*) and the Short-eared Owl, can use their unique auditory powers to locate and seize prey invisible to the eye and hidden in thick vegetation or even under a deep covering of snow [12]. The facial plumage of the owl forms a parabolic dish, or facial disc, edged by a ruff, that focuses and enhances sounds received (**Figure 5**). The ears are located at the sides of the head, behind the eyes, and are covered by the densely packed auricular feathers of the facial disc and ruff (**Figure 6**). The size and shape of the ear opening vary from species to species, with some species also having either a pre-aural or postaural operculum or flap (**Figure 7**).

An owl's range of audible sounds is not unlike that of humans, but its hearing is exceptionally more acute within certain frequencies; particularly at frequencies of 5 kHz and above [13], maximising hunting accuracy with frequencies between 4 and 8 kHz [14]. Some owl species have asymmetrically set ear openings (i.e. one ear is higher than the other). This asymmetry is found in five phyletic lines, represented in the Genera *Tyto, Phodilus, Strix, Rhinoptynx, Asio, Pseudoscops*, and *Aegolius* [15]. Ear asymmetry makes the auditory directional sensitivity pattern for high frequencies different in elevation between the two ears. This allows the owl to localise sound in the vertical plane, by comparing the intensity and spectral composition of sound between the two ears. In simple terms, when a noise is heard, the owl can locate its source because of the minute time difference in which the sound is perceived in the left and right ears. This interaural time difference can be as short as 10 millionths of

**Figure 5.**
*Perfect facial disc of the great Grey. Owl (*Strix nebulosa*). Photo: Tony Hisgett. Source: Wiki commons: https:// creativecommons.org/licenses/by/2.0/deed.en*

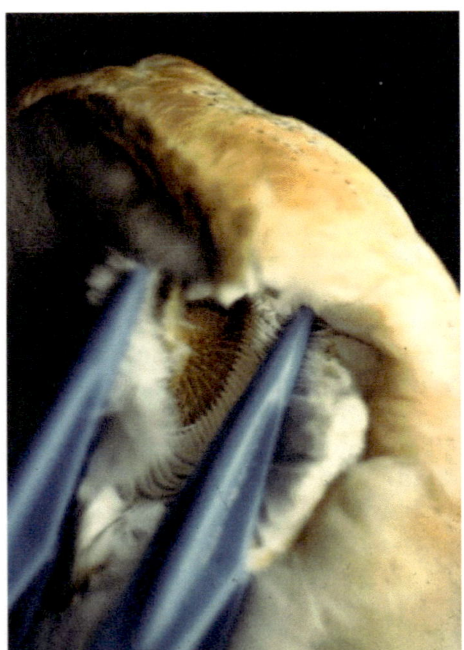

**Figure 6.**
*External ear-opening behind the facial disc of a barn owl (*Tyto alba*). Photo: Alan Sieradzki.*

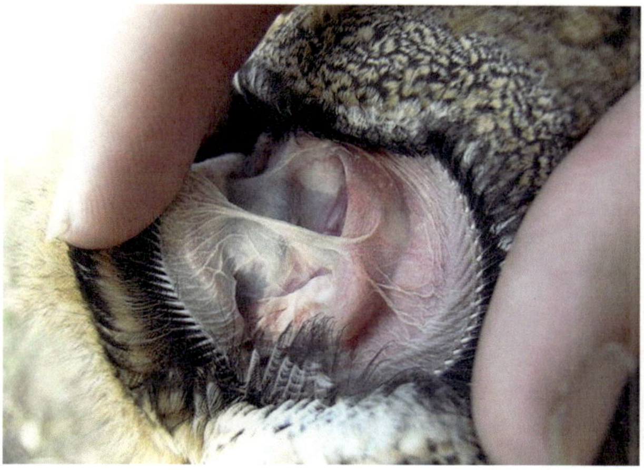

**Figure 7.**
*Operculum of a long-eared owl (*Asio otus*). Photo credit: Creative commons'—https://creativecommons.org/licenses/by-sa/2.5/deed.en*

a second [16]. For example, if the sound was to the left of the owl, the left ear would hear it before the right ear. The owl then turns its head until the sound arrives at both ears simultaneously—the prey, even when not visible due to darkness or cover, is now directly in the owl's line of sight. Even once the prey has been located and the owl has launched an attack, the owl will continue to make minute adjustments of the moveable ruff and flaps until the moment of strike. To understand how this works, research has turned to the area of neurology.

Recent research has discovered that interaural time differences (ITD) are used to localise sounds in azimuth, whereas interaural level differences (ILD) are used to localise sounds in elevation. These two features are processed independently in two separate neural pathways that converge in the external nucleus of the inferior colliculus to form an *auditory map of space* [17]. The brain constructs the auditory space map by comparing the responses of neurons in the two ears to a sound that stimulates both. The left-right positioning of the sound source is computed from the different arrival times of the sound at each ear [18]. Owls with symmetrical ears must determine the horizontal and the vertical directions of a sound separately, one after the other, by tilting head movements [14], thus making it that little bit more difficult a process to lock on to moving prey. Once the prey has been located and locked on to and the owl has launched itself into the attack, movements of the facial ruff and flaps continue to make minute adjustments throughout the flight path until the moment of impact.

## 5. Feathers

Owl feathers are unique in both structure and use. Owls are more heavily feathered than any other bird, even having feathered eyelids and, in many species, feathered feet and toes. In 2017, David H. Johnson (Executive Director of the Global Owl Project) and a small group of volunteers systematically plucked and counted every single feather from the remains of a recently deceased female Great Horned

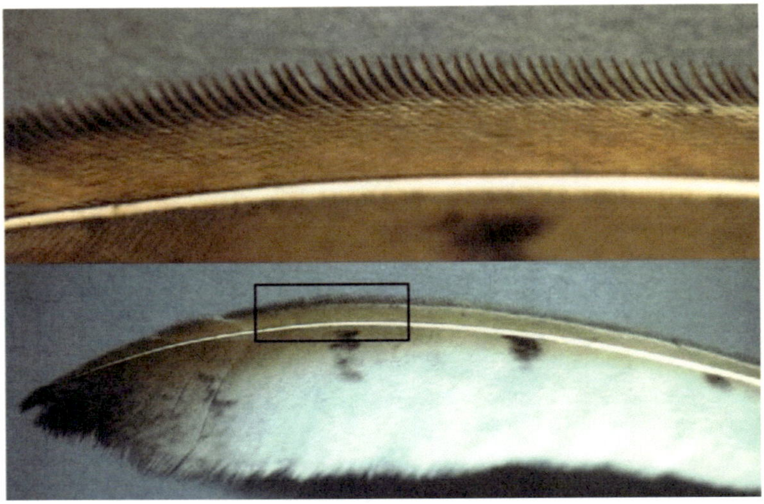

**Figure 8.**
*The primary flight feather of the barn owl (*Tyto alba*) shows the serrated leading edge (fimbriae) and the wispy trailing edge. Photo: Alan Sieradzki.*

Owl (*Bubo virginianus*). This painstaking exercise, taking 46 man-hours of work, resulted in a total count of 12,230 feathers. Full details of this feather count exercise are planned to be published shortly [Johnson, D.H. personal communication].

It is, however, the unique structure and form of the wing feathers, allowing the owl almost silent movement through the air, that are the most remarkable. The owls' near-silent flight can be attributed to three wing feather adaptations unique to owls—(1) a comb-like leading edge to the primary and secondary flight feathers (fimbriae); (2) a fine, wispy, fringe-like trailing edge to the flight feathers (**Figure 8**); and (3) a velvety covering on the upper surface of the wing and a shiny, down-covered underside [19, 20]. The large wings of these birds, resulting in low wing loading (calculated by the weight of the bird divided by the surface area of both wings) [21], and a low aspect ratio, contribute to noise reduction by allowing extremely slow and buoyant flight. Also, the owl's wing feathers can separate from each other in flight, allowing the air to flow over each of the individual flight feathers. With all other birds, air rushes over the surface of the wing creating turbulence which, in turn, produces noise. With an owl's wing, the comb-like serrations on the feather's leading-edge break down the air into little groups of micro-turbulences. This effectively muffles the sound of the air rushing over the wing surface, which is further dampened by the velvety coating on the wing's surface and allows the owl to fly silently [22]. A recent study has shown that there is a direct correlation between the size of the facial disc in relation to the length of the comb-like serrations, suggesting that species that rely more on their auditory system for locating prey also have the more silent flight [23]. This also suggests a dual purpose in the need for silent flight, the need for stealth, allowing the owl to approach prey undetected and the need for self-masking, enabling the owl to locate prey by sound while in flight [23]. Such is the effectiveness of the owl's unique wing feathers for silent flight, that the international aeronautical industry is now investing heavily in the research and development of wing design based on the owl's fimbriae towards solving the aerodynamic noise of aircraft [24].

Other uniquely structured feathers of the owl are their auricular feathers. In almost all owl species, the facial plumage forms a parabolic dish with a facial ruff. The centre of the ruff is formed by tightly packed feathers, with thick rachis and dense webbing. Such feathers are also found on the pre-aural flaps which cover the ear openings, and in the region of the beak. The facial ruff made up of auricular feathers, collect and amplify sounds, and direct them to the ear openings [25]. Three different types of auricular feathers occur in the facial disc of the Barn Owl. One type covers the reflector feathers of the disc and dominates the general appearance of the facial ruff. A similar smaller type of auricular feather is situated at the pre-aural flaps. The third type of auricular feather (semi-bristle) is found in the region of the beak and functions as a mechanoreceptor [26].

## 6. Digestive system

Owls have evolved to eat their smaller prey whole and unlike other birds, they do not have a crop (**Figure 9**). This system reduces the owl's need to drink water, as much of its liquid intake comes directly from the body fluids of its prey. The whole prey is passed head first straight down the oesophagus and into the proventriculus (glandular stomach).

Digestion begins in the proventriculus, which produces digestive enzymes and stomach acid. The food mass, along with the digestive enzymes, then passes into the second part of the stomach, the ventriculous or gizzard (muscular stomach) where the chemical digestion started in the proventriculus continues and manual digestion begins. The gizzard uses strong muscular contractions to aid in digestion. The soft and digestible parts of the food are allowed to continue along with the digestive system into the small intestine [27]. The indigestible parts (fur, feathers, claws, bones etc.) are retained in the gizzard and compacted into an oval-shaped pellet (oval due to the gizzard's shape). The digestion process up to this point takes several hours (**Figure 10**). The pellet is then passed back into the proventriculus where it will remain for several hours more before finally being regurgitated. Additional digestive enzymes are likely digesting any remaining digestible

**Figure 9.**
*Juvenile tawny owl (*Strix aluco*) swallowing rodent prey. Photo: Alan Sieradzki.*

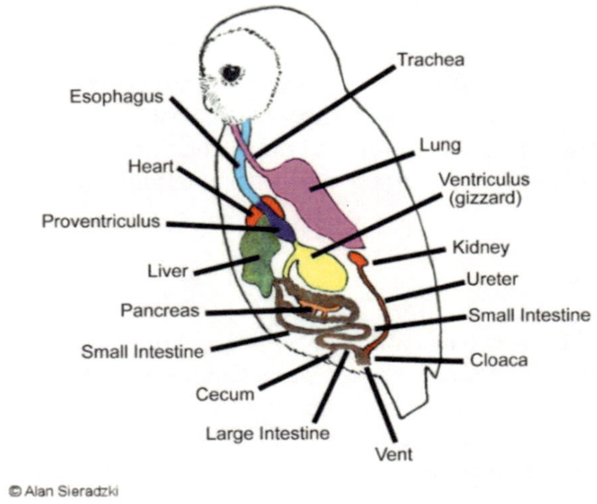


**Figure 10.**
*The digestive system of an owl. Image: Alan Sieradzki.*

**Figure 11.**
*Owl pellet species comparison. L-R: Tawny owl (*Strix aluco*) barn owl (*Tyto alba*), short-eared owl (*Asio flammeus*). Photo: Alan Sieradzki.*

material during this time [28]. Because the stored pellet partially blocks the owl's digestive system, new prey cannot be swallowed until this pellet is ejected.

Regurgitation often signifies that the owl is ready to eat again. When the owl eats more than one prey item within several hours, the various remains are consolidated into one pellet. When the digestive process is finished, the owl will regurgitate the pellet by the process of reverse peristalsis, where smooth muscular contractions push the pellet up the oesophagus and back into the mouth. This process is different from coughing or retching and can prove to be quite strenuous for the owl, especially with larger pellets—this is why an owl will often take on a pained expression when producing a pellet and the reason why owls cannot produce pellets in flight. At the moment of expulsion, the neck is stretched up and forward, the beak is opened, and the pellet simply drops out (**Figure 11**).

## 7. Leg and foot musculature

Owls have developed extremely specialised and powerful musculature in their legs and feet. Contrary to the visual image of an owl at rest, owls have relatively long legs; in some species, they can be as much as half the total body length. In flight, the legs are tucked under the body with the toes closed. Once the prey has been located, however, the owl will swoop down on the prey with its head forward and its feet swinging like a pendulum until the last moment before impact when its head is thrown back and its legs stretched out with its talons open. The eight toes are spread, just before contact, into a symmetrical configuration to cover as large an area as possible [3, 13].

Hind limbs of owls are characterised by the absence of some muscles found in other birds. They lack *m. iliofemoralis, m. ambiens*, both portions of *m. flexor cruris lateralis*, *m. plantaris*, and *m. fibularis longus* [29]. They also have a relatively short tarsometatarsus and the presence of well-developed sesamoids [30] and a specialised tendon locking mechanism [31]. These anomalies in the morphology of the owl's hind limb are associated with adaptations to catch, seize, keep, and kill the large prey [32]. As the owl's normal method of dispatching prey is by impact and constriction (a bite to the neck or skull may also be employed with larger prey), the musculature of the feet and toes are exceptionally powerful. Owls generate more force than Hawks and Falcons when closing their talons, which anatomically translates into stronger digit flexor muscles, more robust bones, and stronger tendons with ossification [28]. Owl's talons are more uniform in size amongst digits, generally less curved, and relatively larger than in diurnal raptors, which probably serve to maintain the reach of the toe for grasping [29].

Owls' feet have extremely thick pads with very prominent papillae (**Figure 12**). Unlike other birds, owls have cone-like papillae, free from one another [33]. The most extreme and specialised papillae are found in the fish-eating owls, where the distance between the papillae is comparatively long and the top sharply pointed [33]. These long-pointed papillae or spicules, help the owl to seize slippery fish and other aquatic prey. The dermal layer in the pad is thick and has a dense structure of collagenous fibres. The dermis functions as a base for the papillae and is the structurally firm part of the skin. The important function of the papillae is to penetrate the roughness of the ground, tree branches, or the fur and skin of prey [34].

Owls are *anisodactylous*, having three toes projecting forward and one toe projecting backwards. However, owls should perhaps be classed as *semi-zygodactylous*, as the outer toe is 'hinged' and can be moved backwards to give the owl's feet two toes projecting forward and two toes projecting backward configuration (*zygodactylous*). This configuration is ideal for perching on branches and seizing prey with cylindrical-shaped bodies (as in rodents) and also allows all four digits to maintain comparable locking power [29]. The anisodactyl configuration allows the owl greater stability on flat surfaces, such as nests, or when subduing larger, struggling prey on the ground. This hinged toe mechanism is not unique to owls as it is shared with the Osprey (*Pandion haliaetus*). The middle toe of the Tytonidae family of owls (Barn, Bay and Grass Owls) have a pectinate talon; a serrated, comb-like flange, used to groom the delicate facial auricular feathers (**Figure 12**). This is another feature shared with the Osprey (*Pandion haliaetus*).

**Figure 12.**
*The foot of a barn owl (*Tyto alba*) shows pads and pectinate talon. Photo: Alan Sieradzki.*

## 8. Variations on a theme

While all owls share these unique adaptations, the evolutionary process of adaptive radiation has produced several variations within the many species [3]. These variations have been influenced by a combination of habitat, prey selection, and activity rhythm (nocturnal, crepuscular, or diurnal).

While all owls share the same ocular morphology, there is a limited variation in iris colour between the various species; either yellow, orange, or black/brown. A recent study has shown that dark eyes are to be found in 71 species belonging to 14 genera, whereas 135 species belonging to 20 genera were classed to have bright eyes (yellow or orange). Dark irises are more frequent amongst strictly nocturnal owls (41 out of 70 nocturnal species [59%]) than amongst owls that have diurnal or crepuscular activity rhythms (37 out of 131 diurnal or crepuscular species [28%]) [35]. The results of the study provided strong support for the existence of an evolutionary correlation between iris colouration and activity rhythm in owls. Beyond that correlation, the study did not find any clear evidence that dark eyes are more likely to evolve in species presenting strictly nocturnal habits than in diurnal species. However, it did find that the most likely explanation for the found patterns would be that dark eyes might be less conspicuous at night and help the owl in avoiding detection by predators or prey.

One of the most distinctive features of the owl is the facial disc (**Figure 5**). However, two groups of owls, Fish Owls and Fishing Owls, have evolved less defined facial discs (also completely lacking the facial disc ruff); almost to the point of being non-existent in the Fish Owl species. There are four species of Fish Owls, the huge Blakiston's Fish Owl (*Bubo blakistoni*), the Brown Fish Owl (*Bubo zeylonensis*) (**Figure 13**), the Tawny Fish Owl (*Bubo flavipes*), and the Malay Fish Owl (*Bubo ketupa*) while there are three species of Fishing Owl, Pel's Fishing Owl (*Bubo peli*), the Rufous Fishing Owl (*Bubo ussheri*), and the Vermiculated Fishing Owl (*Bubo bouvieri*). Fish and Fishing Owls have recently been moved from the Genera *Ketupa* and *Scotopelia* respectively to the Genus *Bubo* in the taxonomic listing because of their shared skeletal and phylogenetic characteristics with Eagle Owls [36].

**Figure 13.**
*Brown fish owl (*Bubo zeylonensis*). Photo: Manojiritty. Source: Creative commons: https://creativecommons.org/licenses/by-sa/4.0/deed.en*

Fish Owls and Fishing Owls are nocturnal and crepuscular hunters and generally search for their aquatic prey from rocks or low hanging tree branches close to the water's edge or wade through the shallow water itself [37]. The less defined facial disc and the lack of the disc ruff (to enhance the acoustic locating of prey) suggest that these characteristics, which are common amongst most other owl species, do not increase the efficiency of hunting aquatic prey visually [37]. As well as lacking the distinctive facial disc found in other owl species, the Fish Owls and the Fishing Owls more or less lack another of the owl's unique adaptations, the comb-like leading edge of the flight feathers (fimbriae) which contributes towards the silent flight in owls [37, 38].

However, it is not just Fish Owls and Fishing Owls that lack the serrations on the leading edge of the flight feathers. A small number of other owl species also lack or have very much less developed fimbriae. These species tend to be primarily diurnal in their activity rhythm [39] and largely insectivorous; species such as the Little Owl (*Athene noctua*), Burrowing Owl (*Athene cunicularia*), Elf Owl (*Micrathene whitneyi*), and Northern Pygmy Owl (*Glaucidium californicum*). Fish and Fishing Owls have no tactical need for silent flight because sound does not travel well between air and water, while the diurnal species similarly have little need for silent flight as they are visible to prey. This would suggest that the vastly reduced fimbriae in these birds are an evolutionary holdover that lacks current function [23].

Further variations can be found in the hind limbs of owls. The extent of feathering on the legs and feet of owls varies from an almost bare tarsus and entirely bare toes to densely long-feathered tarsus and toes. The extent of this variation between species is dictated by geographic location and habitat [40]. An example of this would be to compare the sparsely feathered legs and feet of the grasslands and desert-dwelling Burrowing Owl (**Figure 14**) to the densely feathered legs and feet of the Great Grey Owl of the northern taiga/boreal forests (**Figure 15**).

In 1936, American ornithologist Leon Kelso identified and categorised five types of leg and foot feathering amongst owls, associating each type to a variety of Climatic zones [40]:

**Figure 14.**
*Legs and feet of a burrowing owl (*Athene cunicularia arubensis*). Photo: Global owl project.*

**Figure 15.**
*Densely feathered legs and feet of a great Grey owl (*Strix nebulosa*). Photo: Jari Peltomäki.*

1. *Toes and part of tarsus bare*. Tarsus is bare of feathers all the way around for part or all of its length. Toes entirely bare of feathers—associated with the humid, warm environment of the Tropical, Subtropical, and Temperate Zones; example of species: Cuban Bare-legged Owl (*Gymnoglaux lawrencii*).

2. *Toes bare*. Tarsus fully feathered and at least half of sides and the upper surface of toes bare of feathers—associated with the humid, warm environment of the Tropical, Subtropical, and Temperate Zones; example of species: Tropical Screech-Owl (*Megascops choliba*) and Northern Barred Owl (*Strix varia georgica*).

3. *Toes sparsely feathered or bristled*. Feathers or bristles somewhat thinly distributed over most of upper surface and sides of toes—well represented in all but the colder zones. In the Tropical, Subtropical, and Temperate Zones this type of feathering is more frequently associated with the arid parts of the zones; example of species: Barn Owl and Eastern Screech Owl (*Megascops asio*).

4. *Toes densely short-feathered*. The density of feathering is much greater than in the preceding type, sufficient to hide most of the upper surface of toes from view. Feathers short in comparison to the size of the bird, not tending to conceal part of claws—represented in all the life zones but includes a slightly higher percentage of the owl order in those zones which present a cooler environment, while in zones of greater heat and humidity it constitutes low percentages, example of species: Great Horned Owl (*Megascops asio*) and Short-eared Owl (*Asio flammeus*).

5. *Toes densely long-feathered*. Feathers long in comparison with the size of the bird, tending to conceal part of claws—associated with the colder and less humid environment of the Arctic, Hudsonian, Canadian, and Transition Zones; Example species: Great Grey Owl (*Strix nebulosa*) and Snowy Owl (*Bubo scandiacus*) [39].

Extremes in foot feathering in owls seem to be associated with zones that have extremes of climate and humidity. A perfect example of this is the extremely long and

**Figure 16.**
*The underside of the densely feathered foot of a snowy owl (*Bubo scandiacus*). Photo: Roar Solheim.*

dense feathering of the feet of the Snowy Owl, which gives this ground-nesting owl perfect insulation against the cold Arctic climate and the frozen tundra (**Figure 16**).

The long-legged Burrowing Owl, which lives in an arid climate, has extremely sparse feather covering on its legs and feet, with the density of this covering varying between the subspecies. Generally, however, the female usually has a slightly heavier covering of plumaceous feathers on their upper leg than the male; the reason for this possibility is that the female spends more time than the male in the much cooler environment of the burrow chamber during the nesting season. The Burrowing Owl also has an extra adaptation to its hind limbs, giving it a longer step length and potentially faster limb movements for terrestrial locomotion and possibly for digging [41].

## 9. Shared survival mechanisms and behaviour

While these six unique adaptations combine to make the owl the highly efficient nocturnal predator that it is, the owl also shares several other adaptations and habits which contribute to its survival. Owls have cryptic coloured plumage made up of a mixture of soft browns, greys, black, and white and arranged in subtle markings of streaks and spots which serve to break up the bird's outline, rendering it almost invisible against its background [42]. Some species, such as the Eagle, Owl, and the Long-eared Owl, also have feathered head adornments (ear, tufts, or horns), which help to break up the distinctive round shape of the owl's head [43], while some *Glaucidium* species have developed false eye markings, or an *occipital face* on the nape of their necks, made visible by the fluffing of the head feathers and tilting the head forward [44] (**Figure 17**). Studies have shown that these false eye markings are an effective countermeasure against daytime avian mobbing [45]. Overall, the camouflage of owls is incredibly effective against predators, against being mobbed by other birds during the day, and against being spotted by potential prey at night.

Roughly one third of all owl species are subject to colour polymorphism (colour morph), existing in genera, such as *Strix*, *Tyto*, *Megascops*, *Otus*, *Psiloscops*, *Lophostrix*, *Glaucidium,* and *Bubo*. The pigment melanin is responsible for many observed cases of colour morph, in which there is a great deal of variation within owls and while there are a number of hypotheses surrounding colour morph, the exact mechanisms which drive these variations remain unresolved [46]. One hypothesis, that apostatic selection drives colour morph in owls, where intraspecific colour variation should be promoted in predators by prey forming an avoidance image for the more common colour morph has been proven highly unlikely [47]. A more likely explanation is the niche variation hypothesis, where the species with broader ecological niches should be more variable compared with those with narrow niches because of the action of disruptive selection [48] and that it is an adaptive character likely maintained by the selective advantage of morphs under different environmental conditions via disruptive selection mechanisms [46].

Climate-related colour morph can be seen in the Eastern Screech Owl where individuals exhibit rufous, intermediate, or grey colouration that is likely caused by relative amounts or concentration of black or rufous melanin subtype (eumelanin and pheomelanin, respectively). This species exhibits clinal variation in morph prevalence; the rufous morph predominates in warm climates while the grey morph dominates in a less humid and colder environment (**Figure 18**).

The rufous morph of the Eastern Screech Owl is rarely seen in the northern areas of its range as the mortality rate is greater than that of the grey morph variant in conditions of extreme cold. It is also noted that females of the rufous phase have a greater survival rate in much lower ambient temperatures than their rufous-coloured male counterparts [12]. This greater survival rate in females is probably due to reverse sexual dimorphism (RSD), where the female owl is larger and has more bulk than the male and can capture larger prey.

**Figure 17.**
*The occipital face of Ridgway's pygmy owl (*Glaucidium ridgwayi*). Photo: Bruce Marcot.*

**Figure 18.**
*Two colour morphs of the eastern screech owl. (*Megascops asio*) L – Rufous and R – Grey. Photo: Dick Daniels (http://carolinabirds.org/) https://creativecommons.org/licenses/by-sa/3.0/legalcode*

It can also be hypothesised that colour polymorphism in owls is an adaptive character likely maintained by the selective advantage of camouflage under different light regimes or in terms of physiological adaptation to environmental conditions via disruptive selection mechanisms. Under this hypothesis, climate change could bring about a dramatic change in the colour polymorphism of some northern species. The Tawny Owl (*Strix aluco*) is a colour polymorphic species with a grey and brown morph resident in the Western Palearctic. Studies in Finland have shown that in winter, the grey phase helps to avoid avian mobbing and predators more efficiently than the brown morph and therefore has a higher survival rate in snowy environments. However, as winters are getting milder and shorter in this species range due to climate change, the selection periods promoting grey colouration may eventually disappear [49].

Although some species of owls are specialist feeders, such as Fish Owls and Fishing Owls, and some have a definite preference for certain prey, such as the Barn Owl and the Short-eared Owl with voles, most owls are fairly generalist feeders, with prey as varied as rodents, birds, amphibians, insects and other invertebrates and, in a few opportunistic cases, even bats [50]. Not too long ago, because of their acknowledged diet of live prey, consisting of small vertebrates and invertebrates, it was widely accepted that owls did not scavenge, and any reported observation of this uncharacteristic behaviour was taken as an anomaly. However, recent studies have shown that carrion feeding by owls may be far more prevalent than once thought. In the past, because of their mostly nocturnal activity, dietary information had come mainly from pellet analysis while any observations of scavenging behaviour in owls have been rare and poorly documented. Today, however, with the increasing use of passive infrared wildlife camera traps, baited with a variety of carcasses, including roadkill, a surprising number of owl species have been recorded engaging in this behaviour in Europe, North America, South America, Asia, and Australia (but none, to date, in Africa) [51].

Species recorded scavenging include Barn Owl (*Tyto alba*), Eurasian Eagle Owl (*Bubo bubo*), Tawny Owl (*Strix aluco*), Great Horned Owl (*Bubo virginianus*), Snowy Owl Owl (*Bubo scandiacus*), Ural Owl (*Strix uralensis*), Powerful Owl (*Ninox strenua*), Western Screech Owl (*Megascops kennicotii*), Northern Hawk Owl (*Surnia ulula*), Long-eared Owl (*Asio otus*), Little Owl (*Athene noctua*), and Barred Owl (*Strix varia*). The recorded carrion ranged from feral pigeon to sheep and deer, and there is even a recorded case of a Brown Fish Owl scavenging on the carcass of a crocodile [52–54].

Owls hatch their eggs asynchronously as a survival mechanism against prey shortage. Incubation starts with the laying of the first egg, unlike many other birds that begin incubation with the laying of the last egg of the clutch. With asynchronous hatchings separated by anywhere from a few hours to several days, this gives the older nestlings a distinct advantage in begging for food. British ornithologist David Lack identified asynchronous hatching as an evolutionary adaptation to unpredictable changes in the food supply; if food declines abruptly during the nestling period, the youngest nestlings would die first without endangering the survival of the whole brood [55].

A small number of species, such as the Burrowing Owl, Short-eared Owl, Long-eared Owl and the Eurasian Scops Owl have become seasonally migratory [56]. In North America, Northern Burrowing Owls enter primarily in the southern United States from California to western Louisiana, much of Mexico, and scattered sites southward into Central America [57]. While the soft, lightly oiled feathers of owls are not equipped for long flights over large bodies of water, in Europe, the Eurasian Scops Owl crosses the Mediterranean Sea on its long migration south of the Sahara Desert in Africa, while the Long-eared Owl and the Short-eared Owl regularly fly across the

North Sea from Northern Europe and Scandinavia to winter in the British Isles where they swell the numbers of resident birds [58].

Snowy Owls make nomadic winter movements and are also subject to irruptions; cyclic events triggered by fluctuations in rodent prey population levels [12]. These irruptions can be on such a large scale after a successful breeding season that huge numbers of young owls spread out from the Arctic Circle into southern Canada, Northeast America, and beyond. Although strong fliers, it seems that Snowy Owls are not averse to seeking any advantage in their dispersal. In October 2001, as many as 60 Snowy Owls boarded a ship near Deception Bay, North Quebec, during a severe gale, while a further three landed on another ship east of Newfoundland. Both vessels were heading for the port of Westerscheldt, on the Belgium/Netherlands border. A number of these owls remained on board for the trans-Atlantic crossing, with one individual eventually making it to Felixstowe in England [59].

Owls hitching a ride on ships is nothing new. In 1903, H.W. Henshaw sighted a Short-eared Owl landing on a ship 500 miles (800 km) northwest of the Hawaiian Islands, while in 1901, W.A. Bryan reported a short-eared owl boarding a steamer which plied between Honolulu and Puget Sound, while it was '680 miles off the mainland' [60]. The British Royal Navy has its own bird watching society, the Royal Navy Bird Watching Society (RNBWS), and since its formation in 1946, the RNBWS has been compiling a database of all birds recorded on British Navy vessels around the world. This database includes the names of the vessels and the geographical positions (longitude & latitude). In 2007, Lt. Cdr. Stan Howe, R.N. very kindly extracted a list of all the owls recorded by the RNBWS around the world from its database for me. There were 242 individual sightings listed, which included species, such as Barn Owl (*T. alba*), Eurasian Scops Owl (*Otus scops*), Striated Scops Owl (*Otus brucei*), Collared Scops Owl (*Otus bakkamoena*), Great Horned Owl (*B. virginianus*), Eurasian Eagle Owl (*B. bubo*), Spotted Eagle Owl (*Bubo africanus*), Snowy Owl (*B. scandiacus*), Tawny Owl (*S. aluco*), Barred Owl (*S. varia*), African Wood Owl (*Strix woodfordii*), Northern Hawk Owl (*S. ulula*), Little Owl (*A. noctua*), Spotted Owlet (*Athene brama*), Burrowing Owl (*Athene cunicularia*), Tenmalm's/Boreal Owl (*Aegoleus funereus*), Brown Hawk-owl (*Ninox scutulata*), Moluccan Hawk-owl (*Ninox squamipila*), Jungle Hawk-owl (*Ninox theomacha*), Long-eared Owl (*A. otus*), Short-eared Owl (*Asio flammeus*), and Marsh Owl (*Asio capensis*). The Long-eared Owl (*A. otus*) and Short-eared Owl (*A. flammeus*) are the most recorded species and there are a number of unidentified species listed only as 'Strigidae'. Anyone wishing to access the RNBWS database should visit the RNBWS website: https://www.rnbws.org.uk/science.

With a combination of its six unique adaptations and all these shared survival mechanisms, the owl is indeed one of nature's cleverest survivors.

## Acknowledgements and photo credits

I would like to thank David H. Johnson for allowing me to include the total feather count he and a small group of volunteers painstakingly removed from the remains of a Great Horned Owl (*B. virginianus*) in 2017 before his own publication on the subject and thank Bruce Marcot and Heimo Mikkola for their helpful editorial comments and valuable input. I would also like to thank the following individuals for their kind permission for the use of their photos and images: Deane Lewis (**Figure 2**), Bruce Marcot (**Figure 4**), Jari Peltomäki (**Figure 15**) and Roar Solheim (**Figure 16**).

## Author details

Alan Sieradzki
Global Owl Project, U.S.A.

*Address all correspondence to: naturalistuk@yahoo.co.uk

## References

[1] Espındola-Hernandez P, Mueller JC, Carrete M, Boerno S, Kempenaers B. Genomic evidence for sensorial adaptations to a nocturnal predatory lifestyle in owls. Geonome Biology and Evolution. 2020;**12**(10): 1895-1908. DOI: 10.1093/gbe/evaa166

[2] Walls GL. The Vertebrate Eye and its Adaptive Radiation. Michigan: The Cranbrook Institute of Science; 1942. p. 818

[3] Burton P. What makes an owl. In: Burton JA, editor. Owls of the World. Oxford, England: Peter Lowe (Eurobook Ltd.); 1973. p. 216

[4] Lisney TJ, Iwaniuk AN, Bandet MV, Wylie DR. Eye shape and retinal topography in owls (Aves: Strigiformes). Brain Behavior and Evolution. 2012;**79**:218-236

[5] Sparks J, Soper T. Owls: Their Natural and Unnatural History. Newton Abbot, England: David & Charles; 1973. p. 206

[6] Steinbach MJ, Money KE. Eye movements of the owl. Vision Research. 1973;**13**:889-891

[7] Marek RD, Falkingham PL, Benson RBJ, Gardiner JD, Maddox TW, Bates KT. Evolutionary versatility of the avian neck. Proceedings of the Royal Society B. 2021;**288**:20203150. DOI: 10.1098/rspb.2020.3150

[8] Krings M, Nyakatura JA, Fischer MS, Wagner H. The cervical spine of the American barn owl (*Tyto furcata pratincola*): I anatomy of the vertebrae and regionalization in their S-shaped arrangement. PLoS One. 2014;**9**(3):e91653. DOI: 10.1371/journal.pone.0091653

[9] Grytsyshinaa EE, Kuznetsova AN, Panyutinaa AA. Kinematic constituents of the extreme head turn of *Strix aluco* estimated by means of CT scanning. Doklady Biological Sciences. 2016;**466**:24-27

[10] Krings M, Nyakatura JA, Boumans MLLM, Fischer MS, Wagner H. Barn owls maximize head rotations by a combination of yawing and rolling in functionally diverse regions of the neck. Journal of Anatomy. 2017;**231**(1):12-22. DOI: 10.1111/joa.12616

[11] de Kok-Mercado F, Habib M, Phelps T, Gregg L, Gailloud P. Adaptations of the owl's cervical & cephalic arteries in relation to extreme neck rotation. Posters & Graphics. Science. 2013;**339**(6119):514-515. DOI: 10.1126/science.339.6119

[12] Mikkola H. Owls of the World: A Photographic Guide. 2nd ed. London: Christopher Helm; 2013. p. 528

[13] Payne RS. Acoustic location of prey by barn owls (*Tyto alba*). Journal of Experimental Biology. 1971;**54**:535-573

[14] Knudsen EI, Konishi M. Mechanisms of sound localization in the barn owl (*Tyto alba*). Journal of Comparative Physiology. 1979;**133**:13-21

[15] Norberg RÅ. Independent evolution of outer ear asymmetry among five owl lineages; morphology, function and selection. In: Newton I, Kavanagh R, Olsen J, Taylor I, editors. Ecology and Conservation of Owls. Clayton, Victoria, Australia: CSIRO Publishing; 2002. p. 400

[16] Konishi M. Listening with two ears. Scientific American. 1993;**268**(4):66-73

[17] Kettler L, Griebel H, Ferger R, Wagner H. Combination of interaural level and time difference in azimuthal sound localization in owls. Eneuro. 2017;**4**(6):1-13. DOI: 10.1523/ENEURO.0238-17.2017

[18] Stryker MP. Sensory maps on the move. Science. 1999;**284**(5416):925-926

[19] Graham RR. The silent flight of owls. Journal of the Royal Aeronautical Society. 1934;**38**(286):837-843

[20] Lilley GM. A study of the silent flight of the owl. In: AIAA Paper 1998-2340, 4th AIAA/CEAS Aeroacoustics Conference; 2-4 June 1998; Toulouse, France. Reston, Virginia, U.S.A.: American Institute of Aeronautics and Astronautics; 1998

[21] Johnson DH. Wing loading in 15 species of north American owls. In: Duncan JR, Johnson DH, Nicholls TH, editors. Biology and Conservation of Owls of the Northern Hemisphere. St Paul, MN: US Department of Agriculture; 1997. pp. 553-556

[22] Wagner H, Weger M, Klaas M, Schroder W. Features of owl wings that promote silent flight. Interface Focus. 2017;7:20160078. DOI: 10.1098/rsfs.2016.0078

[23] K Le Piane, CJ Clark. Quiet flight, the leading edge comb, and their ecological correlates in owls (Strigiformes). Biological Journal of the Linnean Society, 2021;**135**(1):84-97. DOI: 10.1093/biolinnean/blab138

[24] Kopania J. Acoustics parameters the wings of various species of owls. In: INTER-NOISE and NOISE-CON Congress and Conference Proceedings, InterNoise16, Hamburg. Reston, Virginia, U.S.A.: Institute of Noise Control Engineering; 2016. pp. 6841-7829

[25] Coles RB, Guppy A. Directional hearing in the barn owl (Tyto alba). Journal of Comparative Physiology A. 1988;**163**:117-133

[26] Wagner H. Morphometry of auricular feathers of barn owls (*Tyto alba*). European Journal of Morphology. 2002;**40**(1):15-21

[27] Grimm RJ, Whitehouse WM. Pellet formation in a great horned owl: A roentgenographic study. The Auk. 1963;**80**:301-306

[28] Smith CR, Richmond ME. Factors influencing pellet egestion and gastric pH in the barn owl. The Wilson Bulletin. 1972;**84**(2):179-186

[29] Volkov SV. The hindlimb musculature of the true owls (Strigidae: Strigiformes): Morphological peculiarities and general adaptations. Ornithologia. 2004;**31**:154-174

[30] Ward AB, Weigl PD, Conroy RM. Functional morphology of raptor hindlimbs: Implications for resource partitioning. The Auk. 2002;**119**(4):1052-1063

[31] Einoder L, Richardson A. The digital tendon locking mechanism of owls: Variation in the structure and arrangement of the mechanism and functional implications. Emu. 2007;**107**:223-230

[32] Fowler DW, Freedman EA, Scannella JB. Predatory functional morphology in raptors: Interdigital variation in talon size is related to prey restraint and immobilisation technique. PLoS One. 2009;**4**(11):e7999

[33] Lennerstedt I. A functional study of papillae and pads in the foot of passerines, parrots, and owls. Zoologica Scripta. 1975;**4**(1):111-123. DOI: 10.1111/j.1463-6409.1975.tb00723.x

[34] Meise W. Zur Systematik der Fischeulen. Ornithologische Monatsberichte. 1933;**41**:169-173

[35] Passarotto A, Parejo D, Cruz-Miralles A, Avilés JM. The evolution of iris colour in relation to nocturnality in owls. Journal of Avian Biology. 2018;**49**(12):1-17. DOI: 10.1111/jav.01908

[36] del Hoyo J, Elliot A, Sargatal J, editors. Handbook of the Birds of the World. Barn-Owls to Hummingbirds. Vol. 5. Barcelona: Lynx Edicions; 1999. p. 759

[37] Hume R, Boyer T. Owls of the World. London: Parkgate Books Ltd.; 1991. p. 192

[38] Slaght JC, Surmach SG, Kisleiko AA. Ecology and conservation of Blakiston's fish owl in Russia. In: Nakamura F, editor. Biodiversity Conservation Using Umbrella Species. Ecological Research Monographs. 2018. pp. 47-70

[39] Weger M, Wagner H. Morphological variations of leading-edge serrations in owls (Strigiformes). PLoS One. 2016;**11**(3):e0149236. DOI: 10.1371/journal.pone.0149236

[40] Kelso L, Kelso EH. The relation of feathering of feet of American owls to humidity of environment and to life zones. The Auk. 1936;**53**(1):51-56

[41] Ilynsky VA. Locomotor adaptations in the hindlimbs of owls: The burrowing owl (*Athene cunicularia*), compared to the little owl (*Athene noctua*). Oryctos. 2008;**7**:271-276

[42] Endler JA. A predator's view of animal color patterns. In: Hecht MK, Steere WC, Wallace B, editors. Evolutionary Biology. Vol. 11. Boston, MA: Springer; 1978

[43] Perrone M Jr. Adaptive significance of ear tufts in owls. Condor. 1981;**83**:383-384

[44] Vesanen M. The occipital face of the pygmy owl Glaucidium passerinum. Ornis Svecica. 2009;**19**:193-198

[45] Deppe C, Holt D, Tewksbury J, Broberg L, Petersen J, Wood K. Effect of northern pygmy-owl (*Glaucidium gnoma*) eyespots on avian mobbing. The Auk. 2003;**120**(3):765-771

[46] Passarotto A, Parejo D, Penteriani V, Avilés JM. Colour polymorphism in owls is linked to light variability. Oecologia. 2018;**187**(1):61-73. DOI: 10.1007/s00442-018-4128-0

[47] Fowlie MK, Krüger O. The evolution of plumage polymorphism in birds of prey and owls: The apostatic selection hypothesis revisited. Journal of Evolutionary Biology. 2003;**16**(4):577-583. DOI: 1046/j.1420-9101.2003.00564x

[48] Galeotti P, Rubolini D. The niche variation hypothesis and the evolution of colour polymorphism in birds: A comparative study of owls, nightjars and raptors. Biological Journal of the Linnean Society. 2004;**82**:237-248

[49] Koskenpato K, Lehikoinen A, Lindstedt C, Karell P. Gray plumage color is more cryptic than brown in snowy landscapes in a resident color polymorphic bird. Ecology and Evolution. 2020;**10**:1751-1761. DOI: 10.1002/ece3.5914

[50] Sieradzki A, Mikkola H. A review of European owls as predators of bats. In: Mikkola H, editor. Owls. London: IntechOpen; 2020. pp. 67-86. DOI: 10.5772/intechopen.80242

[51] Allen ML, Ward MP, Južnič D, Krofel M. Scavenging by owls: A global review and new observations from Europe and North America. Journal of Raptor Research. 2019;**53**(4):410-418

[52] Kapfer JM, Gammon DE, Groves JD. Carrion-feeding by barred owls (*Strix varia*). The Wilson Journal of Ornithology. 2011;**123**(3):646-649

[53] Mori E, Menchetti M, Dartora F. Evidence of carrion consumption behaviour in the long-eared owl *Asio otus* (Linnaeus, 1758) (Aves: Strigiformes: Strigidae). Italian Journal of Zoology. 2014;**81**(3):471-475

[54] Lack D. The Natural Regulation of Animal Numbers. Oxford: Clarendon Press; 1954. p. 343

[55] Mikkola H. Owls of the World Enhanced E-Book. London: Bloomsbury/Christopher Helm; 2014

[56] Melcher CP. Burrowing owl. In: Assal TJ, Melcher CP, Carr NB, editors. Southern Great Plains Rapid Ecoregional Assessment – Pre-Assessment Report: U.S. Geological Survey Open-File Report 2015 – 1003. Reston, Virginia, U.S.A.: U.S. Geological Survey; 2015. p. 284. DOI: 10.3133/ofr20151003

[57] Glue DE. Feeding ecology of the short-eared owl in Britain and Ireland. Bird Study. 1977;**24**(2):70-78. DOI: 10.1080/00063657709476536

[58] Harvey PV, Riddiford N. An uneven sex ratio of migrant long-eared owls, Ringing & Migration. 1990;11(3):132-136. DOI: 10.1080/03078698.1990.9673975

[59] Tomlinson D. BirdOn! News. Jacobi Jayne & Company; 2001. Available from: https://www.birdcare.com/bin/shownews/219

[60] Clark RJ. A field study of the short-eared owl, *Asio flammeus* (Pontoppidan), in North America. Wildlife Monographs. 1975;**47**:3-67

Section 2

# Food Studies

Chapter 2

# What Do We Know about the Diet of Ecuadorian Owls?

*Héctor Cadena-Ortiz, Jorge Brito, María Cristina Ríos, Paolo Piedrahita, Glenda Pozo-Zamora, Hermann Wagner and Juan Freile*

## Abstract

Ecuador territory barely surpasses 28 million hectares, but it is home to 29 owl species, which represent 62% of all owl species in South America. Knowledge of the trophic ecology of owls is fundamental for understanding how they correlate with their environment, which organisms they prey upon, and to which extent they are beneficial for humans' health. Although publications in this field have grown in Ecuador in recent years, background information is still deficient. At present, the diet of a handful of species has been studied in this Andean country, and for a few other species, there is only anecdotal information, such as stomach contents of specimens, held in scientific collections. In this chapter, we review knowledge about the diet of owl species occurring in Ecuador and provide guidelines for the study of pellets with the aim of motivating more research in this field.

**Keywords:** Cricetidae, diet, natural history, owl pellets, prey-predator, Strigiformes

## 1. Introduction

Nocturnal birds of prey birds have long been considered as a "unit" in several classifications of animals [1, 2]. Currently, they are classified in the order Strigiformes, within a larger group called the "African landbird radiation" that includes the orders Cathartiformes, Accipitriformes, Trogoniformes, Coraciiformes, Piciformes, among others [3]. The Strigiformes are difficult to study because they occur in low densities, have generally elusive behaviors, most are strictly nocturnal, and usually remained overlooked [4]. Although research on Strigiformes biology has intensified over the last few decades [2], many owl species remain little studied.

Knowledge about the feeding ecology of Strigiformes, including which organisms are preyed upon them, is a key for understanding their role within a given ecosystem, and even elucidating whether they can be beneficial for human health. However, witnessing owl predation events in the wild is unusual, so analyses of owl pellets become the most effective method for studying their diets. Owls usually swallow their prey whole or in large pieces, and then regurgitate a pellet containing indigestible matter, such as bones, fur, feathers, and other keratinous material, about once a day [5]. Owl pellets are also an efficient alternative for measuring small mammal community composition [6].

Ecuador is home to 29 out of 45 South American species of Strigiformes [7]. For the majority of owls occurring in Ecuador, scarce information about their natural history exists, including their feeding ecology [8]. As there are few reviews of the diet of Neotropical owls at country scales [9, 10], in this chapter, we present a revision of information on the diet of owls in Ecuador. Further, we present some guidelines for the study of owls' diets and highlight topics that could be investigated. Our aim is to promote further scientific research about the feeding ecology of nocturnal birds of prey.

## 2. Previous knowledge

Mammals, birds, reptiles, amphibians, and invertebrates are roughly the usual prey for owls, but what is the preferred prey of each species? There is published information on nine out of the 29 species of owls occurring in Ecuador; the remaining 20 species have only anecdotal data, whereas for two there is virtually no information on their diet in Ecuador. Given that there are few peer-reviewed publications about the diet of Ecuadorian owls, in this section, we also include gray literature, such as conference papers or thesis, and unpublished information compiled by the authors in the field and in a revision of museum specimen labels. The level of taxonomic identification of the prey is variable, according to secondary information. We have tried to identify our own observations to the highest taxonomic level possible.

### 2.1 *Tyto furcata*

*T. f. contempta*: There are only three specimens (MECN 6833, 8773, 8678) of this common and widespread taxon with stomach content information. They contained remains of Coleoptera and rodents. Stomach content of MECN 6992 mentions herbs as the only item, so it cannot be considered secondary content (*i.e.*, stomach contents of a prey); it could have been accidentally ingested when hunting.

There are two studies of its diet using pellets, but from a mammalogy approach, so non-mammal preys were overlooked. These studies underline the high consumption of rodents [11, 12].

This is the best-studied species in Ecuador, most studies being based on pellets analyses, and covering as varied habitats as cities [13, 14], rural areas, or agricultural fields in the Pacific lowlands [15–17] and Andean valleys [18]. However, there is no long-term research on its diet or a broader study aiming to understand feeding preferences in relation to prey abundance and availability.

The predominance of rodents in the diet of *T. furcata* is consistent with studies in other countries [18 and cites therein]. However, this species is also opportunistic, preying upon other kinds of prey, including bats [14] or high number of birds in agricultural and open habitats [18]. Six pellet studies showed the following quantitative data:

In Cuenca, Azuay province, n = 245: 249 introduced rodents, 95 native rodents, 7 birds, and 93 Coleoptera items [13]. In Pallatanga, Chimborazo province, n = 61: 25 introduced rodents (*Mus musculus* and *Rattus* sp.), 132 native rodents, 2 *Sylvilagus andinus*, 3 birds, and 2 reptiles. In Calceta, Manabí province, n = 46: 34 introduced rodent (*M. musculus*), 99 native rodents, 2 *Proechimys decumanus*, 1 *Marmosops* sp., and 1 *Sylvilagus daulensis* [15]. Another study in Cuenca, Azuay province, n = 32: 100 introduced rodents, 29 native rodents, 1 vampire bat *Desmodus rotundus*, 2 birds, 4 reptiles, and 18 Coleoptera [14]. In Zapotillo, Loja province, n = 65: 130 *M. musculus* and 42 native rodents [16]. In another sample from Calceta, Manabí province, n = 30: 57 native rodents, 7 bats, and 1 dove [17]. The largest pellet sample studied in Ecuador included 361 pellets and 664 g of pellets debris from San Antonio and Tababela, Pichincha province. This sample included 86 introduced rodents, 803 native rodents, 67 *S. andinus*, 93 birds, 6 reptiles, and 63 insects [18].

*T. f. punctatissima:* The Galapagos endemic subspecies is little studied as compared to other Galapagos land birds [19]. The first available information was provided from a sample of 1217 pellets collected in Santa Cruz and Isabela islands [20]. It reports 2230 mammal prey (mostly *M. musculus*), 81 birds, 10 reptiles, and 2440 invertebrates. A second study reported 90% of rodents, 14% of insects, and less than 1% of birds from a sample of 104 pellets collected on Santa Cruz Island [21, 22]. Details of specific prey found in the latter study will be published elsewhere.

## 2.2 *Megascops albogularis*

*M. a. albigularis:* Stomach contents of three museum specimens (MECN 6150, 7755, 7760) contained remains of a myriapod, also orthoptera, and other insects [23].

There is no other available information on its diet, but one wild pair is being fed raw chicken pieces in Zuro Loma, Pichincha province (**Figure 1**). The behavioral implications of this artificial feeding are not known, but there are records of carrion consumption by owls, including *Megascops* species [24]. This event might allow close observations of this owl by birders and nature lovers, which could increase their empathy and appreciation, and promote its conservation. This case deserves further study.

*M. a. macabrum*: Stomach contents of one museum specimen (MECN 772) had insect remains [23].

## 2.3 *Megascops choliba crucigerus*

Stomach contents of one museum specimen (MECN 3994) had insect remains [23]. Another museum specimen (MECN 9633), not included in the cited work, also had insect remains.

## 2.4 *Megascops koepckeae*

An unreported number of pellets of this species was only recently reported for Ecuador [25], collected in Loja city, and contained mainly insects and the introduced *Mus musculus* [26]. There is one record of nestling being provisioned with one scorpion and one frog [D. Pacheco pers. obs.].

## 2.5 *Megascops ingens*

*M. i. ingens:* Stomach contents of five museum specimens (MECN 773, 7288, 7758, 7763, 7773) contained remains of the order Orthoptera (crickets, grasshoppers) and other unidentified insects [23].

*M. i. colombianus:* One individual mist-netted in Río Guajalito, Pichincha province, carried an opossum *Marmosops impavidus* on its talons, of which the entire head and forelimbs had already been eaten [27]. There is an additional observation of Carabidae (Coleoptera) remains in an abandoned nest, in Nanegalito, Pichincha province (JF. Freile pers. obs.).

**Figure 1.**
Megascops albogularis *feeding on raw chicken pieces in a private reserve in Pichincha. Photo: Courtesy of Edison Buenaño.*

## 2.6 *Megascops petersoni*

Stomach contents of six museum specimens (MECN 774, 775, 7752, 7762, 7866, 7867) had remains of the order Orthoptera (crickets, grasshoppers) and other unidentified insects [23].

### 2.7 *Megascops centralis*

Stomach contents of three museum specimens (FMNH 372109, MECN 5994, 7062) had remains of an Orthoptera, Coleoptera, other insects, and two centipedes [23]. These authors erroneously reported caterpillar remains of specimen MECN 6361, which is actually *Megascops roboratus* (see below). Another museum specimen (MECN 9987) had insect remains.

Direct observations in a nest at Loma Alta, Santa Elena province, where recorded five food delivery events to nestlings [28]. These authors also collected three pellets, some pellet debris, and prey remains in the nest. Prey provisioned to nestlings included 55 Tettigoniidae, Acrididae, Coleoptera, Lycosidae, Amblypygi, Anura, and Passeriformes. No quantitative data on each prey are presented. There is an observation of a direct attack on a cicada in Mashpi, Pichincha province (JF. Freile pers. obs.).

### 2.8 *Megascops roraimae napensis*

Stomach contents of three museum specimens (MECN 7004, 7636, 7772) had remains of the orders Coleoptera, Orthoptera, other insects, and a bone [23]. MECN 8183 also had remains of insects.

### 2.9 *M. roboratus*

*M. r. pacificus:* Stomach contents of one museum specimen (MECN 771) had insect remains [23]; another specimen (MECN 6301) had caterpillars.

There is no information on the diet of *M. r. roboratus*.

### 2.10 *Megascops watsonii*

Stomach contents of three museum specimens (ANSP 186787, MECN 969, 7764) had remains of spiders, also Orthoptera, and other insects [23].

### 2.11 *Lophostrix cristata*

*L. c. wedeli:* Stomach contents of one museum specimen (MECN 6444) had remains of Orthoptera and other unidentified insects [23].

No information on the diet of *L. c. cristata*.

### 2.12 *Pulsatrix perspicillata*

*P. p. perspicillata:* One pellet from Cuyabeno, Sucumbíos province, had one lizard *Thecadactylus solimoensis* [29].

*P. p. chapmani:* Stomach contents of one museum specimen (QCAZ 4503) had two *Iguana* juveniles and a moth larva [23].

In a sample of nine pellets collected in Zapotillo, Loja province, 12 mammal preys were found, along with one frog, one reptile, and one invertebrate. The native rodent *Rhipidomys leucodactylus* was the most important prey in terms of frequency (40%) and in biomass contribution (54%) [30].

**Figure 2.**
Pulsatrix melanota *recently fledged juvenile and its pellet in the inset, July 2019. Photos: MC. Ríos and courtesy of Luis Gualavisí.*

### 2.13 *Pulsatrix melanota*

Stomach contents of three museum specimens (QCAZ 3508, MECN 6838, 7774) contained only insects of the orders Hemiptera, Mantodea, Orthoptera, and Phasmatodea [31]. Additionally, some pellets of a recently fledged juvenile observed in Tundayme, Zamora Chinchipe province (**Figure 2**) had Coleoptera remains, mainly (MC. Ríos unpubl.). An adult stayed close to the juvenile, while the observer approached it, and she was even attacked by the adult.

### 2.14 *Bubo virginianus nigrescens*

*Sylvilagus andinus* were the main prey brought to a nest in Cotopaxi volcano, Cotopaxi province [32]. There is an additional observation of an adult preying upon a *S. andinus* in Antisana, Napo province, and a rat, presumably *Rattus* sp., in Cañón del Chiche, Pichincha province (JF. Freile pers. obs.).

### 2.15 *Strix virgata*

*S. v. virgata:* Stomach contents of three museum specimens (ANSP 181031, MECN 6991, 8350) contained remains of Coleoptera, other insects, and one reptile [23]. No information from the Amazonian population, whose subspecific identity remains unsolved [33].

### 2.16 *Strix nigrolineata*

The only available documentation is a video of an adult preying upon a moth (Sphingidae) in Buenaventura Reserve, El Oro province (https://macaulaylibrary.org/

asset/201777541). It has been observed capturing large moths (undetermined families) in light posts at Mindo, Pichincha province, and Los Cedros, Imbabura province (JF. Freile, pers. obs.).

### 2.17 *Strix huhula*

There is one unpublished observation of hunting bats in a streetlight in Puyo, Pastaza province (JF. Freile pers. obs.). In addition, a photographic record of one adult preying upon a rat in San Isidro, Napo province (https://macaulaylibrary.org/asset/457375971), pertains to an isolated population whose subspecific identity remains unresolved [34]. There are several additional observations from San Isidro of large Lepidoptera and Coleoptera predation at streetlights, as well as unidentified bats and one unidentified rodent [Holroyd & Trefry unpubl.].

### 2.18 *Strix albitarsis*

One small mammal without a skull for identification and four Coleoptera of the genus *Megaceras* were found in three pellets collected in Yanayacu, Napo province [35]. Stomach contents of four museum specimens (MECN 793, 6927[ex 927], 6135, QCAZ 1544) contained remains of insects in the orders Blattodea and Coleoptera, other unidentified insects, and rodent fur [23].

### 2.19 *Glaucidium nubicola*

Stomach contents of one museum specimen had insect remains and one lizard (ANSP 181044 [holotype]) and another specimen had insects (ANSP 180178) [36].

There is an observation of an adult feeding a juvenile with a lizard [37]. There is also one photographic record of an adult preying upon the lizard *Andinosaura oculata* in Mindo, Pichincha, at 17 h00 [38]. These authors erroneously identified the observed owl as *Glaucidium jardinii*, which occurs at higher elevations.

One individual ringed in Reserva Las Tangaras, Pichincha province, had one lizard *Pholidobolus vertebralis* on its talons when mist-netted and when recaptured [39]. Terrestrial lizards apparently represent an important feature of this owl diet (JF. Freile unpubl.).

### 2.20 *G. jardinii*

Stomach contents of five museum specimens (LSUMZ 112509, 112,510, MECN 787, 6034, 7868) had remains of a Coleoptera, other insects, one rodent, and one mammal [23].

### 2.21 *Glaucidium parkeri*

Stomach contents of one museum specimen (ANSP 185160) had rests of a bird [23].

### 2.22 *Glaucidium griseiceps*

No data from Ecuador and its diet elsewhere are also poorly known, but possibly include insects, spiders, and small vertebrates [40].

### 2.23 *Glaucidium brasilianum ucayalae*

Stomach contents of four museum specimens (QCAZ 1452, ANSP 186790, FMNH 316441, MECN 783) had Coleoptera and other insects' remains [23].

### 2.24 *Glaucidium peruanum*

Stomach contents of four museum specimens (MECN 3921 [ex 392], 6134, 6302; QCAZ 3658, LSUMZ 77569) contained remains of the orders Coleoptera, Odonata, Orthoptera, and other unidentified insects [23]. Additionally, there is a photographic record of an adult with a *Holcosus septemlineatus* lizard on its talons, taken in Uzcurrumi, El Oro province (Bravo X., in litt, **Figure 3**). Another photographic record, taken in Puerto López, Manabí province, involves the geckoo *Phyllodactylus reissii* [41]. This observation was made at night. There is an additional observation of an adult preying upon a *Microlophus occipitalis* lizard in Zapotillo, Loja province (JF. Freile pers. obs.).

### 2.25 *Athene cunicularia*

*A. c. pichinchae:* There are only three specimens (MECN 8754, 8755, 8756) of this common and widespread taxon with stomach contents information, which are only insects. There is only one study of its diet, in a sample of 40 pellets from Piedra Labrada, Loja province, was found 84 rodent prey, 1 reptile, 7 frogs, 593 insects, and 187 other invertebrates. Insects were the most important prey in terms of frequency (80%) and rodents in biomass contribution (95%) [42]. In another sample of 368 pellets from Tababela, Pichincha province, was found 13 rodent preys, 2 reptiles, 156 insects, and 25 other invertebrates. Insects were the most important prey in terms of frequency (68%) and rodents in biomass contribution (75%) [42].

*A. c. punensis:* Stomach contents of two museum specimens (MECN 779, 6303) had remains of Orthoptera [23]. There is also an observation of predation on frog *Rhinella marina* in Macará, Loja (JF. Freile pers. obs.).

All studies available for this taxon are consistent with insects as the most important prey in frequency and rodents in biomass contribution [16, 43–45]. The largest pellet sample studied included 300 pellets from Atahualpa, Santa Elena province; it reported 1981 invertebrates, 266 mammals, 10 reptiles, and 3 birds [43]. This work reports two *G. peruanum* as prey, based on bills in the pellets, but there are no further details. There are no previous reports of intra-guild predation by *A. cunicularia*, and it remains plausible that the bill could have been misidentified.

Another study from Jambelí, El Oro province, reported 929 insects, 165 other invertebrates, 151 introduced rodents, and 19 birds, in 182 pellets [44]. Further, in 48 pellets collected in Zapotillo, Loja province, 503 insects, 9 frogs (Bufonidae), 9 wolf spiders (Lycosidae), and 5 scorpions were included [16]. Lastly, a sample of 50 pellets from Calceta, Manabí province, included 544 insects and 16 rodents as preys [45].

No information about subspecies *A. c. carrikeri*, which is apparently spreading along river islands in the Amazon lowlands [33].

### 2.26 *Aegolius harrisii harrisii*

There is no diet information for this subspecies. König et al. [40] mention small vertebrates and insects as probable prey.

**Figure 3.**
Glaucidium peruanum *preying upon a lizard* Holcosus septemlineatus *in El Oro, May 2019, 14 h30. Photo: Courtesy of Xavier Bravo Guerrero.*

## 2.27 *Asio clamator clamator*

Stomach contents of one museum specimen (QCAZ 1413) from Bahía de Caráquez, Manabí province, had an opossum *Marmosa simonsi* [23].

A study of 72 pellets and 284.4 g of pellet debris from Atahualpa, Santa Elena province, also included 1 *M. simonsi*, 351 rodents, 12 birds, 310 reptiles, 3 frogs

*Ceratophrys stolzmanni*, and 468 insects [46]; insects were the most important prey in terms of frequency (41%) and reptiles in biomass contribution (58%). Authors reported seasonal differences in diet with greater consumption of mammals and reptiles during the dry season, and of insects in the wet season.

Another study is in progress in Cuenca, Azuay province, found principally rodents in its pellets (H. Cadena-Ortiz et al. unpubl.). This study is in an unusual locality for the owl, because there are scarce records from Andean valleys and cities [33].

### 2.28 *Asio stygius robustus*

Cadena-Ortiz et al. [47] studied 38 pellets, 127 g of pellet debris and seven prey remains from Quito, Pichincha province, and reported 136 birds, 1 bat, and 20 Coleoptera. The dove *Zenaida auriculata* was the most important prey in terms of frequency (74%) and biomass contribution (91%). There is an additional observation of predation on *Z. auriculata* in Tumbaco, Pichincha province (JF. Freile pers. obs.).

### 2.29 *Asio flammeus*

*Asio f. bogotensis:* 55 rodent preys, 30 other mammals, 8 birds, 1 reptile, 1 frog, and 72 Coleoptera were found in 52 pellets in three locations of Pichincha province [48]. Mammals were the most important prey in terms of frequency (51%) and biomass contribution (87%). Another study of 163 pellets collected in Antisana, Napo province, included 112 rodent preys, 111 *Sylvilagus andinus*, and 19 Coleoptera. *S. andinus* was the most important prey in terms of frequency (46%) and biomass contribution (78%) [49].

*A. f. galapagoensis:* Only five publications exist on the diet of this Galapagos endemic subspecies. First by de Groot [20], who sampled 213 pellets from six islands (Santa Cruz, Champion, Española, Genovesa, Pitt, and Plaza) and found 58 introduced rodents, 320 birds, and 32 invertebrates. A recent study of 45 pellets reported 47% rodents, 37% birds, and 15% insects [21, 22]. Details of specific prey found in the latter study will be published elsewhere. There is also a report of predation of the Galapagos marine iguana *Amblyrhynchus cristatus* from Isabela Island [50] and the species is known to prey upon seabird nestlings, including storm-petrels [51].

## 3. Overview

Degree of knowledge about the diet of owls in Ecuador can be ranked into four categories: null (no information), scarce (less than five events reported), limited (at least five events reported), and moderate (at least two studies in different localities). There is no extensive rank in Ecuador, and it will be an owl with many studies over the long term in various habitats and many variables. We define five events as a minimal sample for statistical tests; an event is a museum skin, a pellet, a photograph, an observation, or a report of an independent predation event (**Table 1**).

In order to roughly illustrate the diet breadth of each owl taxa in Ecuador, we organized overall prey into five groups (mammals, birds, reptiles, amphibians, and invertebrates). We excluded studies with an exclusive mammalogical approach or studies without quantitative data on prey (**Table 1**).

| Owl | Prey | | | | | Source | | | Level of knowledge |
|---|---|---|---|---|---|---|---|---|---|
| | Mammals | Birds | Reptile | Amphibians | Invertebrates | Pellets | Museum skin | Predation record | |
| *Tyto furcata contempta* | 1962 | 106 | 12 | | 174 | 840 | 3 | | 3 |
| *Tyto furcata punctatissima* | 2230 | 81 | 10 | | 2440 | 1217 | | | 3 |
| *Megascops albogularis albigularis* | | | | | x | | 3 | 1 | 1 |
| *Megascops albogularis macabrum* | | | | | x | | 1 | | 1 |
| *Megascops choliba crucigerus* | | | | | x | | 2 | | 1 |
| *Megascops koepckeae koepckeae* | x | | | 1 | x | x | | x | 1 |
| *Megascops ingens ingens* | | | | | x | | 5 | | 2 |
| *Megascops ingens colombianus* | 1 | | | | x | | | 2 | 1 |
| *Megascops petersoni* | | | | | x | | 6 | | 2 |
| *Megascops centralis* | | x | | x | x | 3 | 4 | 6 | 2 |
| *Megascops roraimae napensis* | | | | | x | | 4 | | 1 |
| *Megascops roboratus pacificus* | | | | | x | | 2 | | 1 |
| *Megascops roboratus roboratus* | | | | | | | | | 0 |
| *Megascops watsonii watsonii* | | | | | x | | 3 | | 1 |
| *Lophostrix cristata wedeli* | | | | | x | | 1 | | 1 |
| *Lophostrix cristata cristata* | | | | | | | | | 0 |
| *Pulsatrix perspicillata perspicillata* | | | 1 | | | 1 | | | 1 |
| *Pulsatrix perspicillata chapmani* | 12 | | 3 | 1 | 2 | 9 | 1 | | 2 |
| *Pulsatrix melanota melanota* | | | | | x | 1 | 3 | | 1 |
| *Bubo virginianus nigrescens* | x | | | | | | | x | 1 |

| Owl | Prey | | | | | Source | | | Level of knowledge |
|---|---|---|---|---|---|---|---|---|---|
| | Mammals | Birds | Reptile | Amphibians | Invertebrates | Pellets | Museum skin | Predation record | |
| *Strix virgata virgata* | | | 1 | | x | | 3 | | 1 |
| *Strix virgata superciliaris* | | | | | | | | | 0 |
| *Strix nigrolineata* | | | | | x | | | x | 1 |
| *Strix huhula huhula* | x | | | | x | | | x | 1 |
| *Strix albitarsis* | 1 | | | | 4 | 3 | 4 | | 2 |
| *Glaucidium nubicola* | | | 7 | | x | | 2 | 6 | 2 |
| *Glaucidium jardinii* | 2 | | | | x | | 5 | | 2 |
| *Glaucidium parkeri* | | 1 | | | | | 1 | | 1 |
| *Glaucidium griseiceps* | | | | | | | | | 0 |
| *Glaucidium brasilianum ucayalae* | | | | | x | | 4 | | 1 |
| *Glaucidium peruanum* | | | 3 | | x | | 4 | 3 | 2 |
| *Athene cunicularia pichinchae* | 97 | | 3 | 7 | 961 | 408 | 3 | | 3 |
| *Athene cunicularia punensis* | 433 | 22 | 10 | 10 | 4136 | 580 | 2 | | 3 |
| *Athene cunicularia carrikeri* | | | | | | | | | 0 |
| *Aegolius harrisii* | | | | | | | | | 0 |
| *Asio clamator clamator* | 388 | 14 | 310 | 3 | 469 | 17 | 1 | | 3 |
| *Asio stygius robustus* | 1 | 137 | | | 20 | 38 | | 1 | 2 |
| *Asio flammeus bogotensis* | 308 | 8 | 1 | 1 | 91 | 215 | | | 3 |
| *Asio flammeus galapagoensis* | 58 | 320 | 1 | | 32 | 213 | | 1 | 3 |

**Table 1.**
*Subspecies of owls present in Ecuador. The numbers of prey indicate consumed individuals, x when the work is merely qualitative. Knowledge level ranges from 0 for null to 3 for moderate.*

Although this exercise considers prey at a very coarse level (classes), it allows us to see some trends for each owl taxa. In owls with moderate information, there is consumption of items in each class, even if in low numbers, which reinforces the opportunistic behavior of the owls. Additionally, the level of knowledge is only moderate for seven owls in Ecuador. Although nine owl taxa present limited knowledge, only *Pulsatrix perspicillata chapmani, Strix albitarsis, Glaucidium nubicola*, and *Asio stygius robustus* have quantitative data. Then *G. nubicola* could indicate that it is specialized only in reptiles, particularly in lizards. Nonetheless, there are also unquantified records of insect consumption which might suggest that *G. nubicola* has a tendency to consume lizards but is not specialized in them. On the other hand, owl taxa with larger samples are similar trends with other works, as has already been suggested for *Athene cunicularia* specialization in insects [44 and cites therein], *T. furcata* in rodents [18 and cites therein], and *A. stygius robustus* in birds [47 and cites therein].

## 4. How to study owl diets

Documenting prey capture by owls in the wild is exceptional unless a nest is found—although the specific identity of prey provisioned to nestlings cannot always be determined. Therefore, the study of pellets is pivotal for understanding species' diets given the ease of analyzing them and the number of samples that can be obtained with relatively little field effort, once roosting and feeding sites are located.

Following known procedures in pellet studies [9], we suggest following these steps to maximize data collection: 1) Remain at the site of pellet collection until species identity can be confirmed. Alternatively, camera traps can be set up for species documentation and identification. 2) Collect pellets individually and also all pellet debris. 3) Georeference the collection site and make a description of the area. 4) Air dry each pellet for at least 3 days, then measure its maximum length and width with a caliper and weigh it. 5) Soak the pellets individually in water and disaggregate them up to separate bones and other prey remains. From our experience in Ecuador, only skulls, jaws, beaks, mandibles, and elytra are useful for identification purposes. 6) Compare your voucher specimens directly with museum specimens. 7) Use the presence of unique structures, skulls, and pairs of mandibles or elytra to estimate the number of prey per pellet as the minimum number of individuals (MNI). 8) Calculate the percentage of occurrence as the MNI of each species by the total number of individuals of all species. 9) Calculate the biomass consumption as the mean body mass (in g) of each species multiplied by its MNI.

To calculate dietary niche breadth, we suggest standardized Levin's index [52], which varies from 0 (narrow trophic niche, maximum prey selectivity) to 1 (wide niche, minimum selectivity). Thus, when the values are less than 0.6, the organism is considered a specialist, since it uses a low number of resources and has a preference for certain foods [53, 54]. To determine owls' foraging strategy, it is necessary to study potential prey richness and abundance in the study area. Strategies could be opportunistic when it ingests the prey in the same relative abundances of its environment, or selective, when it ingests some or all of the prey in different proportions to those present in the hunting area [54]. Degree of dietary overlap between areas or seasons can be analyzed using Pianka's dietary niche overlap index or through a Chi-square test ($\chi^2$) to check for differences in the diet composition between sites or seasons.

## 5. Relevance to mammalogy

Owl pellets can be an effective alternative for measuring small mammal community composition over large geographic areas due to the relative ease and low cost of field collections [6]. Since it is a noninvasive indirect tool, it allows the collection of valuable osteological information. In Ecuador, analysis of pellets has allowed the recording of rare and difficult-to-collect mammal species like *Ichthyomys hydrobates* found in western Ecuador 26 years after the last documented record [15]. Pellet contents have also provided an approximation of the species richness of small mammals in areas with no previous information. For example, we have the first data on presence and abundance of native rodents in previously unstudied Ecuadorian localities in the Andes [49] and western lowlands [15].

Monitoring invasive species (*M. musculus*, *Rattus rattus*, *R. novergicus*) by means of pellets could be a mid- and long-term strategy. They are now known to be agricultural pests because they devastate crops, damage the soil, or eat stored agricultural products [55]. They are also a public health problem because contaminate human food with their excrement [56]. Further, they have a severe impact on several endemic and native species preying on or competing for sources [57]. *Rattus* has caused the extinction of birds on islands, as well as reptiles, small mammals, amphibians, invertebrates, and plants [58]. They are a latent threat to the human species, due to the number of viruses and bacteria they can transmit in their feces, urine, or by direct contact through bites [58].

## 6. Conclusions

Specific records of a predation event, whether in a stomach content or in field observation, could indicate preference or opportunism of an individual owl. Records of more prey, such as those obtained by pellets, could indicate local or temporal preferences of a population. The diet of owls is possibly the best-known aspect of their natural history, due in part to analyses of their pellets [40]. To date, we have a broad idea of prey selection by few species in Ecuador, but no idea about predator-prey interactions for any species. For example, we do not have studies of trophic ecology for any species of owl in Ecuador, developed over the long term in various habitats throughout its distribution. These studies analyze parameters, such as prey availability in the environment and the selectivity of the predator [9]. Most information published comes from anecdotal observations or from studies limited temporally and spatially, and we are still documenting the general aspects of natural history and distribution, but not yet assessing patterns and processes [34].

Rescue centers and/or zoos are the potential sources of relevant information that is currently being lost. Many recently captured or rescued owls arrive at these centers, and some might have prey in their stomachs, providing information on diet and digestion times [59], and in this way not only rescue individuals but also contribute to adding nature history information and eventually in real support for the conservation of the species. However, bureaucratic pitfalls and lack of trained personnel in the national environmental authority discourage research, also they do not store complete or useful information about their rescues, many rescued species are misidentified, and often several "rescues" are not actually needed since they are juveniles that recently abandoned their nest and are learning to fly; rescuing them results in taking them away from their parents and habitats.

It is important to continue publishing natural history reports, including, for example, visual or photographic records of specific predation events that can shed light on owls' capture methods or activity schedules. To date, most publications about Ecuadorian owls pertain to *T. furcata*, mirroring the situation of this species in other Neotropical countries [9]. More information on pellets comes from dry areas such as the southwestern tropics and the Andean valleys since pellets are better preserved in dry than in humid environments like the Amazonian rainforests or cloud forests along Andean slopes. Dietary information of humid forest owls reported for Ecuador to date corresponds to less than three pellets [29, 35].

An owl species' diet is often inferred from its bill form and/or knowledge about its congeners. Yet, diets may vary due to factors, such as seasonality, sex, ontogeny, availability of prey, or geography, even in nearby or similar geographical areas, individuals could differ in diet [60]. Therefore, it is important to continue studying and publishing information even about common species like *T. furcata* and *A. cunicularia* (**Figure 4**). On the other hand, since *Megascops roboratus* and *G. nubicola* are regional endemics (i.e., their global distribution ranges are mainly confined to Ecuador [34]), studying them in Ecuador is promising as to obtain natural history information.

**Figure 4.**
Athene cunicularia *preying upon a Coleoptera in Tungurahua. Photo: Courtesy of Christiana Fattorelli.*

Of 29 species of owls present in Ecuador, there are no diet data in the country for two species (*Glaucidium griseiceps* and *Aegolius harrisii*). If we narrow this analysis to the 39 subspecies present, four additional taxa are added to this figure: *Megascops roboratus roboratus, Lophostrix cristata cristata, Strix virgata* cf. *superciliaris*, and *Athene cunicularia carrikeri*. There are only 11 museum specimens with information about stomach contents deposited in Ecuadorian bird collections, all from MECN, that were not studied earlier [23]. Most information in specimen labels, though, is basic and unquantified. Likewise, there are only eight new papers [14, 18, 28, 30, 44, 46, 47, 49] reporting owl species diets since the only state-of-the-art revision by Freile et al. [34].

## Acknowledgements

Thanks to Edison Buenaño (www.swordbilledexpeditions.com), Luis Gualavisí, Xavier Bravo Guerrero, and Christiana Fattorelli for sharing photos for this chapter, and to curators of natural history museums for sharing or letting us review information: Cesar Garzón, Museo Ecuatoriano de Ciencias Naturales del Instituto Nacional de Biodiversidad (MECN-INABIO); Santiago Burneo, Pontificia Universidad Católica del Ecuador (QCAZ-PUCE); Edith Montalvo, Escuela Politécnica Nacional (EPN); Leonardo Ordoñez, Universidad Técnica Particular de Loja, Bernarda Vásquez, Universidad del Azuay; and Félix Man-Ging, Universidad de Guayaquil.

## Conflict of interest

The authors declare no conflict of interest.

## Author details

Héctor Cadena-Ortiz[1,2]*, Jorge Brito[1], María Cristina Ríos[2], Paolo Piedrahita[3], Glenda Pozo-Zamora[1,2], Hermann Wagner[4] and Juan Freile[5]

1 Instituto Nacional de Biodiversidad (INABIO), Quito, Ecuador

2 Pajareando Ando Ecuador, Ecuador

3 Facultad de Ciencias de La Vida, Escuela Superior Politécnica Del Litoral, Guayaquil, Ecuador

4 Institute of Biology II, RWTH Aachen University, Aachen, Germany

5 Comité Ecuatoriano de Registros Ornitológicos (CERO), Ecuador

*Address all correspondence to: fercho_cada@yahoo.es

## References

[1] Bruce M. A brief history of classifying birds. In: del Hoyo J, Elliott A, Christie DA, editors. Handbook of the Birds of the World. Barcelona: Lynx Edicions; 2003;**8**:11-43

[2] Wink M, Sauer-Gurth H. Molecular taxonomy and systematics of owls (Strigiformes): An update. Airo. 2021;**29**:487-500

[3] del Hoyo J. All the Birds of the World. Barcelona: Lynx Edicions; 2020

[4] Mikkola HJ. Diversity of the owl species in the Amazon region. Ecosystem and biodiversity of Amazonia. In: Ecosystem and Biodiversity of Amazonia. London: InTech Open; 2020. pp. 107-119. DOI: 10.5772/intechopen.94977

[5] Marti CD. Food consumption and pellet formation rates in four owl species. The. Wilson Bulletin. 1973:178-181

[6] Heisler LM, Somers CM, Poulin RG. Owl pellets: A more effective alternative to conventional trapping for broad-scale studies of small mammal communities. Methods in Ecology and Evolution. 2016;**7**(1):96-103

[7] Remsen JV Jr, Areta JI, Bonaccorso E, Claramunt S, Jaramillo A, Lane DF, Pacheco JF, Robbins MB, Stiles FG, Zimmer KJ. A classification of the bird species of South America. American Ornithological Society. Version [24 July 2022]. 2022. Available from: http://www.museum.lsu.edu/~Remsen/SACCBaseline.htm. [Accessed: September 12, 2022]

[8] Freile JF, Castro DF, Varela S. Estado del conocimiento, distribución y conservación de aves rapaces nocturnas en Ecuador. Ornitología Neotropical. 2012;**23**:235-244

[9] Pardiñas UF, Cirignoli S. Bibliografía comentada sobre los análisis de egagrópilas de aves rapaces en Argentina. Ornitología Neotropical. 2002;**13**:31-59

[10] Jaksic F. Ecología de los vertebrados de Chile. Santiago: Ediciones Universidad Católica de Chile; 1997. p. 262

[11] Moreno P. Mamíferos presentes en la dieta de la lechuza de campanario (*Tyto alba*) en Valdivia, provincia de Guayas, Ecuador. ACI Avances en Ciencias e Ingenierías. 2010;**3**:B87-B90

[12] Moreno P, Román JL. Clasificación del género *Reithrodontomys* en el Ecuador y comentarios sobre la alimentación de la lechuza de campanario (*Tyto alba*) en los alrededores de Quito. Boletín Técnico, Serie Zoológica. 2013;**89**:1623

[13] Charpentier AL, Martínez DJ. Abundancia y dieta de *Tyto alba*, Lechuza de Campanario, en Cuenca. Cuenca: Universidad del Azuay; 2007

[14] Vásquez Ávila BE, Nivelo-Villavicencio CH, Picón Renteria PR, Armijos MK, Vásquez CB, Astudillo X. La lechuza Campanaria, *Tyto alba* (Strigiformes: Tytonidae) como regulador de plagas en un ecosistema urbano altoandino en el sur del Ecuador. ACI Avances en Ciencias e Ingenierías. 2018; **10**(16):42-51. DOI: 10.18272/aci.v10i1.975

[15] Brito J, Orellana-Vásquez H, Cadena-Ortiz H, Vargas R, Pozo-Zamora G, Curay J. Mamíferos pequeños en la dieta de la lechuza *Tyto alba* (Strigiformes: Tytonidae) en dos localidades del occidente de Ecuador, con ampliación distribucional de *Ichthyomys hydrobates* (Rodentia: Cricetidae). Papéis Avulsos de Zoologia (São Paulo). 2015;**55**(19):261-268. DOI: 10.1590/00311049.2015.55.19

[16] Cabrera D. Análisis de la dieta de la lechuza de campanario (*Tyto alba*) y el búho terrestre (*Athene cunicularia*) en Zapotillo Ecuador. Loja: Universidad Técnica Particular de Loja; 2018

[17] Estay S, San AP. Potencial de la lechuza blanca (*Tyto furcata*) como controlador biológico en el Campus Politécnico de la Espam "MFL" y sus alrededores. Calceta: Escuela Superior Politécnica Agropecuaria de Manabí Manuel Félix López; 2019

[18] Cadena-Ortiz H, Pozo-Zamora GM, Brito J, Barriocanal C. Diet of barn owls (*Tyto alba*) in two Ecuadorian dry forest locations. Ornitología Colombiana. 2019;**17**:eNB03

[19] Freile JF, Carrión JM, Prieto-Albuja F, Suárez L, Ortiz-Crespo F. La ornitología en Ecuador: Un análisis del estado actual del conocimiento y sugerencias para prioridades de investigación. Ornitología Neotropical. 2006;**17**:183-202

[20] de Groot RS. Origin, status and ecology of the owls in Galápagos (Ecuador). Ardea. 1983;**71**:167182

[21] Wagner H, Bairlein F, Piedrahita P. Foraging of Barn Owls on Santa Cruz (Galapagos, Ecuador): a preliminary study. In: Abstracts of the 149th Jahresversammlung der Deutschen OrnithologenGesellschaft. Stralsund; 2016. p. 23

[22] Piedrahita P, Albán K, Quezada G, Lara D, Wagner H. Composición de la dieta y áreas de caza de la lechuza de campanario, *Tyto alba punctatissima*, en la isla Santa Cruz. Simposio 60 años: Conservación y ciencia en Galápagos. Puerto Ayora, Galápagos. 2019

[23] Cadena-Ortiz H, Freile JF, Bahamonde-Vinueza D. Información sobre la dieta de algunos búhos (Strigidae) del Ecuador. Ornitología Neotropical. 2013;**24**(4):469-474

[24] Allen ML, Ward MP, Južnič D, Krofel M. Scavenging by owls: A global review and new observations from Europe and North America. Journal of Raptor Research. 2019;**53**(4):410-418. DOI: 10.3356/0892101653.4.410

[25] Ordónez-Delgado L, Freile J. First records of Koepcke's ScreechOwl *Megascops koepckeae* (Aves: Strigidae) in Ecuador. Revista Ecuatoriana de Ornitología. 2019;**5**:25-29. DOI: 10.18272/reo.vi5.1193

[26] Ordóñez-Delgado L, Orihuela-Torres A, Freile J. Notas sobre la historia natural del autillo de Koepcke *Megascops koepckeae* (Strigidae) en Ecuador. In: Freile J, Guevara E, CisnerosHeredia DF, Amigo X, Santander T. Memorias de la VI Reunión Ecuatoriana de Ornitología. Revista Ecuatoriana de Ornitología. 2019;**4**:162

[27] Freile JF, Chaves JA. Photospot: Colombian screech-owl,Otus ingens colombianus. Cotinga. 1999;**12**:95-96

[28] Reyes EMR, Astudillo-Sánchez E. Notes on the Nest, owlets, diet, and parasites of the Choco ScreechOwl (*Megascops guatemalae centralis*) in Loma Alta communal reserve, Western Ecuador. The Wilson Journal of Ornithology. 2017;**129**(2):377381. DOI: 10.1676/16019.1

[29] Daza JD, Price LB, Schalk CM, Bauer AM, Borman AR, Peterhans JK. Predation on southern Turniptailed geckos (*Thecadactylus solimoensis*) by a spectacled owl (*Pulsatrix perspicillata*). Cuadernos de Herpetología. 2017;**31**(1):37-39

[30] Orihuela-Torres A, Ordóñez-Delgado L, Verdezoto-Celi A, Brito J. Diet of the spectacled owl (*Pulsatrix*

*perspicillata*) in Zapotillo, southwestern Ecuador. Revista Brasileira de Ornitologia. 2018;**26**(1):52-56

[31] Cadena-Ortiz H, Bahamonde-Vinueza D, Bonaccorso E. Notas sobre la dieta del Búho Ventribandeado (*Pulsatrix melanota*) en Ecuador. Ornitología Neotropical. 2011;**22**:471-475

[32] de Vries T. Presas y periodo de reproducción del Cuscungo (*Bubo virginianus*) en el páramo del Cotopaxi, Ecuador. In: Memorias de las V Jornadas Ecuatorianas de Biología. Sociedad Ecuatoriana de Biología. Quito; 1981. pp. 34-35

[33] Freile J, Restall R. Birds of Ecuador. London: Helm Field Guides; 2018

[34] Freile JF, Guevara EA, Pacheco C, Santander T. The owls of Ecuador. In: Enriquez P, editor. Neotropical Owls. Cham: Springer; 2017. pp. 373-395

[35] Greeney HF. Brief observations on the diet, day roost, and juveniles of the rufous- banded owl (*Strix albitarsis*) in eastern Ecuador. Lundiana. 2003;**41**:67-68

[36] Robbins MB, Stiles FG. A new species of PygmyOwl (Strigidae: Glaucidium) from the Pacific slope of the northern Andes. The Auk. 1999;**116**:305-315. DOI: 10.2307/4089365

[37] Greeney HF, Nunnery T. Notes on the breeding of northwest Ecuadorian birds. Bulletin of the British Ornithologists' Club. 2006;**126**:38-45

[38] Dueñas MR, Valencia JH. *Andinosaura oculata* (Tropical Lightbulb Lizard). Herpetological Review. 2019;**50**:134. DOI: 10.47051/NDQZ5372

[39] Re B, Henry CS, Becker CD. Notes on the natural history of Cloudforest PygmyOwls (*Glaucidium nubicola*) in Ecuador. The Wilson Journal of Ornithology. 2021;**133**(2):314-318. DOI: 10.1676/2000045

[40] König C, Weick F, Becking JH. Owls of the World. 2nd ed. London: Christopher Helm; 2008

[41] Allan-Miranda NA, Ramírez-Jaramillo SM. Predación de *Glaucidium peruanum* (Strigiformes: Strigidae) sobre *Phyllodactylus reissii* (Squamata: Phyllodactylidae), Puerto López,Ecuador. The Biologist (Lima). 2018;**16**(1):181-184. DOI: 10.24039/rtb2018161231

[42] Cadena-Ortiz H, Garzón C, Villamarín-Cortéz S, Pozo-Zamora G, Echeverría-Vaca G, Yánez J, et al. Diet of the burrowing owl *Athene cunicularia*, in two locations of the interAndean valley Ecuador. Revista Brasileira de Ornitología. 2016;**24**:122-128. DOI: 10.1007/BF03544340

[43] Reyes ER. Abundancia Relativa y Dieta del búho terrestre *Athene cunicularia punensis* (Chapman, 1914) en las zonas circundantes de la comuna Atahualpa, Provincia de Santa Elena, Ecuador [thesis]. Guayaquil: Universidad de Guayaquil; 2015

[44] Orihuela-Torres A, OrdóñezDelgado L, Brito J, López F, Mazón M, Freile JF. Ecología trófica del búho terrestre *Athene cunicularia punensis* en el archipiélago de Jambelí, provincia de El Oro, suroeste de Ecuador. Revista Peruana de Biología. 2018;**25**(2):123-130. DOI: 10.15381/rpb.v25i2.13376

[45] Pérez D, Zambrano J. Potencial de la lechuza pequeña (*Athene cunicularia*) como controlador biológico en el Campus Politécnico de la Espam "MFL" y sus alrededores. Calceta: Escuela Superior Politécnica Agropecuaria de Manabí Manuel Félix López; 2019

[46] Vargas R, Abella J, Gregori M, Brito J. Análisis de egagrópilas del búho listado,

*Asio clamator*, (Aves: Strigiformes) en Atahualpa, provincia de Santa Elena, Ecuador. Huitzil Revista Mexicana de Ornitología. 2021;**22**(2):e625. DOI: 10.28947/hrmo.2021.22.2.551

[47] Cadena-Ortiz H, Bedoya J, Pozo-Zamora GM, Watson J, Brito J. Notas sobre la dieta, desarrollo de jóvenes y distribución del Búho Estigio *Asio stygius* (Strigiformes: Strigidae) en Ecuador. Revista Ecuatoriana de Ornitología. 2018;**3**:20-31. DOI: 10.18272/reo.v0i3.773

[48] Pozo-Zamora G, Brito J, García R, Alarcón I, CadenaOrtiz H. Primeras observaciones de la dieta del Búho Orejicorto *Asio flammeus bogotensis* (Strigiformes: Strigidae) en Pichincha, Ecuador. Revista Ecuatoriana de Ornitología. 2017;**1**:1-7. DOI: 10.18272/reo.v0i1.463

[49] Cadena-Ortiz H, Solórzano MF, Noboa M, Brito J. Diet of the short eared owl (*Asio flammeus*) in the Antisana highlands, Ecuador. Huitzil. 2019;**20**(2):e535. DOI: 10.28947/hrmo.2019.20.2.436

[50] Iudica CA, Andrade LF, Sheffield SR. Galapagos ShortEared owl (*Asio flammeus galapagoensis*) preying upon a marine Iguana (*Amblyrhynchus cristatus*) on Isla Isabela, Galapagos Islands, Ecuador. Journal of Raptor Research. 2021;**55**(1):124-126. DOI: 10.3356/0892101655.1.124

[51] Jackson MH. Galapagos, a Natural History. Calgary: University of Calgary Press; 1993

[52] Marti CD. Raptor food habits studies. In: Pendleton BAG, Milsap BA, Cline KW, Bird DM, editors. Raptor Management Techniques Manual. Washington D.C: National Wildlife Federation; 1987. pp. 67-80

[53] Krebs CJ. Species diversity measures. In: Ecological Methodology. 2nd ed. Menlo Park: AddisonWelsey Educational Publishers; 1999. pp. 410-454

[54] Jaksic FM. What do carnivorous predators cue in on: Size or abundance of mammalian prey? A crucial test in California, Chile, and Spain. Revista Chilena de Historia Natural. 1989;**62**:237-249

[55] Witmer G. Rodents in agriculture: A broad perspective. Agronomy. 2022;**12**:1458. DOI: 10.3390/agronomy12061458

[56] Meerburg B, Singleton G, Kijlstra A. Rodent-borne diseases and their risk to public health. Critical Reviews in Microbiology. 2009;**35**:221-270. DOI: 10.1080/10408410902989837

[57] Banks PB, Hughes NK. A review of the evidence for potential impacts of black rats (*Rattus rattus*) on wildlife and humans in Australia. Wildlife Research. 2012;**39**(1):78-88. DOI: 10.1071/WR11086

[58] Denys C, Taylor PJ, Aplin KP. Family Muridae. In: Wilson DE, Lacher TE Jr, Mittermeier RA, editors. Handbook of the Mammals of the World. Vol. 7. Rodents II. Barcelona: Lynx Edicions; 2017. pp. 536-597

[59] Janzen DH, Pond MC. Food and Feeding Behavior of a Captive Costa Rican Least Pygmy Owl *Glaucidium minutissimum Rarum* Griscom. Vol. 9. Aves: Brenesia; 1974. pp. 71-80

[60] Bolnick DI, Svanbäck R, Fordyce JA, Yang LH, Davis JM, Hulsey CD, et al. The ecology of individuals: Incidence and implications of individual specialization. The American Naturalist. 2003;**161**(1):128. DOI: 10.2307/3078879

Section 3

# Voice Studies

Chapter 3

# The Vocal Activity of Twelve African Owl Species

*Heimo Mikkola and Anita Mikkola*

## Abstract

Vocalization of different species of owls carries a lot of scientific information on their distribution and diversity. There is little information on the owl vocalizations in the tropical environments. The calling of 12 African owl species was studied in Malawi 1993–1998, and in The Gambia 1998–2004. The direct listening method was used to collect some 2062+ vocal records mostly at the house gardens or sometimes on the balcony at wildlife lodges and rural hotels. Owls are normally vocal most actively especially just before breeding, but in this material, the peak months coincided very little with the given breeding times in Malawi and The Gambia. It is possible that the validity of the breeding times is not enough, but other reasons are not clear because the sampling was opportunistic rather than systematic. If vocalizations are used to estimate owl populations, it is important to identify the environmental factors affecting owl calling. Heavy rain and wind were silencing the owls or at least made it impossible to hear their voices due to the background noise. Barn Owls *Tyto alba* were often calling immediately after the heavy wind calmed and no precipitation was falling. The temperature is not so important if the other conditions are suitable for calling, Barn Owl and Pearl-spotted Owl *Glaucidium perlatum* were vocal as well in +37°C as in +15°C. Some owl species may increase vocal displays during full moonlight (like the African Barred Owlet *Taenioglaux capense*, Pearl-spotted Owl, and Southern White-faced Owl *Ptilopsis granti*) but others call less or not at all during the full moon (Barn Owl). The impact of the full moon was not that obvious as the bright sky can also activate the Pearl-spotted Owl. Barn Owl started to call actively again when the moon was diminishing to 60% of its full size and its luminosity. It was noted in The Gambia that the obvious predation risk and interference competition was altering the vocal activity of the African Scops Owl *Otus senegalensis*, which stopped calling when the Barn Owl was active. Barn Owl is a predator that can attack the smaller African Scops Owl. In Malawi Spotted Eagle Owl, *Bubo africanus* calling bouts were suspected to silence the Pearl-spotted and Southern White-faced Owls as the larger owl could prey on these smaller owls if hearing their calls.

**Keywords:** owl vocalization, daily activity, seasonality, Strigiformes, Malawi, The Gambia

## 1. Introduction

Most owl species have nocturnal or crepuscular habits, and due to darkness, owl communication relies greatly on vocal activity. In temperate regions, it is the owls which in late winter or early spring fill the night with music, and in the tropics, owls are just part of a formidable chorus of animal songs and calls. Calls are completely diagnostic of species, and owls are likely to recognize other individuals by voice as by sight during their travels in the dark [1, 2]. Every vocalization in an owl's vocabulary has a precise meaning in the communication with conspecifics. Calling is advertising their presence, to locate and attract potential mates, and to establish or to reaffirm breeding territories [3].

In Africa, some owls call almost daily at sunset or soon thereafter, and others are vocal in the still hours before dawn. Often their calling bouts are only momentary and fragmentary, but sometimes, especially just before breeding, they are loud, complex, and prolonged, extending almost throughout the night [4]. Calling at dusk may be mandatory—at least in species that reside in their territories throughout the year—as a notification to neighbors that the owners are still in residence. Later calling may be timed to take advantage of good conditions for sound transmission, or it is a response to social pressure such as neighbors calling, intruders present in the territory, or prospective mates being noticed [4].

The chapter tries to tackle the following questions: (a) What are the main factors influencing the temporal patterns (daily and seasonal) of vocal behavior of African owls; (b) how some environmental variables, like moon, wind, and rain, will influence the calls for some species, and (c) if and how useful vocal records are in surveying the distribution and population size of the different owls. At the same time, this chapter will also give some anecdotes collected on the human nightlife and the nocturnal behavior of some domestic animals and wildlife near the house.

Nowadays, it starts to be outdated and old fashion to collect vocal activities of any owls just by listening and writing down the results as the use of automated bird presence recognition is becoming the modern method for wildlife monitoring. It is felt to be more beneficial for avian biodiversity conservation [5].

This could be the last opportunity to put on record these old African owl call studies from Malawi and The Gambia between 1993 and 2004.

## 2. Material and methods

The vocal activity of 12 African owl species was recorded in Malawi 1993–1998, and in The Gambia 1998–2004. The main species studied were as follows:

Barn Owl *Tyto alba (Malawi and The Gambia).*

Pearl-Spotted Owl *Glaucidium perlatum (Malawi and The Gambia).*

Southern White-faced Owl *Ptilopsis granti (Malawi).*

Northern White-faced Owl *Ptilopsis leucotis (The Gambia).*

Spotted Eagle Owl *Bubo africanus (Malawi).*

To a lesser extent also African Barred Owlet *Taenioglaux capense*, African Scops Owl *Otus senegalensis*, Eurasian Scop Owl *Otus scops*, Greyish Eagle Owl *Bubo cinerascens*, Milky Eagle Owl *Bubo lacteus* and Pel's Fishing Owl *Bubo peli* have been studied. And a few literature references are given on African Wood Owls *Strix woodfordii* that was rarely heard only in Malawi. Summary of the species studied and the number of observations are listed in **Table 1**.

Direct listening was the method used and mostly at the house gardens or sometimes at lodges and hotels on the balcony. Like most Africans, many evenings and even night hours were spent outside the house but sometimes a good TV program was disturbing the study as was heavy rain and strong wind. Similarly, the annual leave spent in

May–June in Finland explains a low number of observations in those months. Field notes included time, weather especially if something unusual in wind, rain and temperature, and the visibility and phase of the moon.

| Countries/Species | Malawi(M) Number of calls | The Gambia(G) Number of calls | Total calls recorded | Remarks |
|---|---|---|---|---|
| African Barred Owl(ABO) | 20 | 0 | 20 | Occurs only in The Gambia |
| African Scops Owl (ASO) | 10 | 20 | 30 | |
| African Wood Owl (AWO) | 5 | 0 | 5 | Rare in The Gambia |
| Barn Owl (BO) | 222 | 504 | 726 | |
| Eurasian Scops Owl (ESO) | 0 | 30 | 30 | Rare in Malawi |
| Giant or Milky Eagle Owl (MEO) | 10 | 2 | 12 | |
| Greyish Eagle Owl (GEO) | 0 | + | + | Gambia Sound recordings exist |
| Northern White-faced Owl (NwfO) | 0 | 62 | 62 | Occurs only in The Gambia |
| Pel's Fishing Owl (PFO) | + | 10 | 10 | Occurs but not heard in Malawi |
| Pearl-spotted Owl (PSO) | 743 | 58 | 801 | |
| Spotted Eagle Owl (SEO) | 54 | 0 | 54 | Occurs only in Malawi |
| Southern White-faced Owl (SwfO) | 312 | 0 | 312 | Occurs only in Malawi |
| *Other animals:* | | | | |
| Mosquito | regular | regular | nr | Cft! |
| Termites | seasonal | rare | nr | Cft! |
| Hyena | regular | 0 | na | Not heard in The Gambia |
| Dogs | regular | regular | na | Cft! |
| Rooster | regular | regular | na | Cft! |
| Spitting Cobra | rare | 0 | Nr | Not seen in The Gambia |
| *Human activities:* | | | | |
| African drums | regular | regular | na | Cft! |
| Muslim prayers | 0 | regular | na | Heard only in The Gambia |
| Heavy shooting | regular | 0 | na | Heard only in Malawi |
| **Total of calls** | 1376+ | 686+ | 2062+ | |

*Other animal species and some disturbing human activities are also recorded. In the brackets, all abbreviations are used in the text when presenting the results. In this table: No = Number; na = not available; nr = not relevant; and Cft = Cf the text!*

**Table 1.**
*Summary of studied 12 owl and number of calls recorded in Malawi 1993–1998 and the Gambia 1998–2004.*

Similarity index

In **Tables 2** and **3**, the similarity index has been calculated to show if there are noticeable time and seasonal differences in the calling activities between Malawi and The Gambia. The index used is modified from MacNaughton & Wolf's [6] "Index on Community Similarity":

$$\text{Similarity Index} = \Sigma(2\,\text{m}) / \Sigma(\text{a} + \text{b})$$

| **Owl** | **BO** | **BO** | **PSO** | **PSO** | **SwfO** | **NwfO** | **SEO** |
|---|---|---|---|---|---|---|---|
| **Country/Hours** | **Malawi** | **Gambia** | **Malawi** | **Gambia** | **Malawi** | **Gambia** | **Malawi** |
| 00:00–01:00 | 10.4 | 11.5 | 3.5 | 3.4 | 4.2 | — | — |
| 01:00–02:00 | 5.9 | 5.2 | 1.7 | — | 0.6 | 3.2 | 13.0 |
| 02:00–03:00 | 9.9 | 3.0 | 1.5 | 6.9 | 1.9 | 11.3 | — |
| 03:00–04:00 | 5.4 | 2.2 | 0.6 | 6.9 | 3.5 | 12.9 | 5.6 |
| 04:00–05:00 | 4.9 | 4.4 | 1.5 | 6.9 | 5.8 | 14.5 | 3.7 |
| 05:00–06:00 | 0.9 | 3.8 | — | 13.8 | — | 14.5 | 3.7 |
| 06:00–07:00 | — | 1.0 | 1.1 | 10.3 | — | 6.5 | — |
| 07:00–08:00 | — | — | 0.9 | 3.4 | — | — | — |
| 08:00–09:00 | — | — | 0.3 | — | — | — | — |
| 09:00–10:00 | — | — | 0.4 | — | — | — | — |
| 10:00–11:00 | — | 0.4 | — | — | — | — | — |
| 11:00–12:00 | — | — | 1.3 | — | — | — | — |
| 12:00–13:00 | — | — | 0.5 | — | — | — | — |
| 13:00–14:00 | — | — | 0.9 | — | — | — | — |
| 14:00–15:00 | — | — | 0.1 | — | — | — | — |
| 15:00–16:00 | — | — | 0.6 | — | — | — | — |
| 16:00–17:00 | — | — | 1.1 | 6.9 | — | — | — |
| 17:00–18:00 | — | — | 5.1 | — | — | — | — |
| 18:00–19:00 | 18.0 | — | 5.5 | 12.1 | 8.4 | — | 33.3 |
| 19:00–20:00 | 9.0 | 11.3 | 7.4 | 5.3 | 20.6 | 4.8 | 13.0 |
| 20:00–21:00 | 5.0 | 18.6 | 13.5 | 1.7 | 11.6 | 14.5 | 11.1 |
| 21:00–22:00 | 9.9 | 15.5 | 21.0 | 10.3 | 19.6 | 9.7 | 7.4 |
| 22:00–23:00 | 12.2 | 13.1 | 20.5 | 3.5 | 19.3 | — | 5.6 |
| 23:00–24:00 | 8.6 | 10.1 | 11.0 | 8.6 | 4.5 | 8.1 | 3.7 |
| Total | 100.1 | 100.1 | 100.0 | 100.0 | 100.0 | 100.0 | 100.1 |
| No of calls | 222 | 504 | 743 | 58 | 312 | 62 | 54 |
| Similarity Index | 0.34 | | 0.26 | | 0.18 | | |

*BO = Barn Owl; PSO = Pearl-spotted Owl; SwfO = Southern White-faced. Owl; NwfO = Northern White-faced Owl, and SEO = Spotted Eagle Owl.*

**Table 2.**
*Timing of the African owl calls as hourly percentage from the total calls.*

| Owl | BO | BO | PSO | PSO | SwfO | NwfO | SEO |
|---|---|---|---|---|---|---|---|
| **Country/Month** | **Malawi** | **Gambia** | **Malawi** | **Gambia** | **Malawi** | **Gambia** | **Malawi** |
| January | 1.4 | 7.5 | — | 10.3 | — | 13.1 | — |
| February | — | 3.6 | — | 1.7 | — | 8.2 | — |
| March | 2.2 | 0.8 | — | 1.7 | — | 9.8 | 58.8 |
| April | 25.2 | — | — | 12.1 | 1.0 | — | — |
| May | 19.8 | 3.0 | — | — | — | — | — |
| June | — | 0.4 | — | — | 0.6 | — | — |
| July | 9.0 | 29.9 | 0.6 | — | 0.6 | — | — |
| August | 6.8 | 18.2 | 0.9 | 3.4 | 2.6 | — | 2.9 |
| September | 4.9 | 8.7 | 5.6 | 3.5 | — | 13.1 | — |
| October | 9.0 | 5.9 | 29.9 | 20.7 | — | 8.2 | — |
| November | 18.5 | 14.1 | 62.9 | 43.1 | 95.2 | 45.9 | 32.4 |
| December | 3.2 | 7.9 | 0.1 | 3.5 | — | 1.7 | 5.9 |
| Total | 100.0 | 100.0 | 100.0 | 100.0 | 100.0 | 100.0 | 100.0 |
| No of calls | 222 | 504 | 743 | 58 | 312 | 62 | 54 |
| Similarity Index | 0.42 | | 0.21 | | 0.05 | | |

*BO = Barn Owl; PSO = Pearl-spotted Owl; SwfO = Southern White-faced. Owl; NwfO = Northern White-faced Owl, and SEO = Spotted Eagle Owl.*

**Table 3.**
*Timing of the African owl calls as monthly percentages from total calls.*

in which: a = percentage in Malawi, b = percentage in The Gambia, m = minimum percentage in either country. The nearer 1.0 the index is, the higher is the similarity between the countries and less important are the noted differences. Due to the non-systematic nature of material collection, no further statistical tests were undertaken.

## 3. Results

### 3.1 Timing and seasonality of the calls

**Table 2** shows the hourly timing of 1955 calls listed for Barn Owls; Pearl-spotted Owls; Southern White-faced Owls; Northern White-faced Owls; and Spotted-Eagle Owls. Most of the Barn Owl calls (18% of all calls) in Malawi were heard between 18:00–19:00 h. and in The Gambia between 20:00–21:00 h. (18.6%). Daytime calling was recorded only in The Gambia. The similarity index between the timing of the Gambian and Malawian Barn Owls was low 0.34 (**Table 2**).

Pearl-spotted Owl's most active calling took place between 21.00 and 22:00 h. in Malawi (21%) and 18:00–19:00 in the Gambia (12%). In the much larger material from Malawi, also the daytime calls were recorded almost every hour of the day, while in The Gambia, there was a real break in the vocal activity between 08:00–16:00 (**Table 2**). Lack of daytime calls in The Gambia may explain the low similarity index between the two countries, 0.26.

In Malawi, the Southern White-faced Owl had the most active calling time 19:00–20:00 h. (20.6% of all calls), and no call was recorded between 05:00–18:00. In The Gambia, Northern White-faced Owl was most active in the small material between 03:00–06:00 with 42% of all calls. The similarity index between The Gambian and Malawian White-faced Owls is very low (0.18) (**Table 2**). This could be partly due to the generous size difference in the materials but could also support the species separation.

Spotted Eagle Owl material from Malawi was much larger but every night callings AM was hearing when preparing the dinner in the kitchen were not written down daily. However, even all noted calls show that Spotted Eagle Owl is the most active vocally between 18:00–19:00 h. (33% of all calls). No daytime calls were recorded between 05:00 and 18:00 h. (**Table 2**).

The seasonality of the vocal activity on the same owl species is presented in **Table 3**. Barn Owl was most vocal in April and November in Malawi and in July, August, and November in The Gambia but calls were heard all year round in both countries, especially when noting that June was not sampled due to the absence in the country. The breeding season of the Barn Owl in Malawi is May–November and September–April in The Gambia [7]. The similarity index between The Gambian and Malawian Barn Owls was a little higher in the months (0.42) than it was in the hours (0.34).

In Malawi and The Gambia, Pearl-spotted Owl was vocally active mainly in October and November, in Malawi (93% of all calls) and in The Gambia (64% of all calls). Interestingly, these months are not coinciding at all with the later presented breeding seasons for this owl neither in Malawi (August–September) nor in The Gambia (February–April) [7]. The similarity index was almost the same in the months as it was in the hours (0.21 vs. 0.26) (**Tables 2** and **3**).

In Malawi, the Southern White-faced Owl was heard almost entirely in November (95%), again not coinciding with the peak breeding season stated to be August [7]. In The Gambia, the Northern White-faced Owl had in this small material two clear peaks, one in January and the other in September and November (**Table 3**)—once more these months are not well in line with the listed breeding seasons [7]. The similarity index was almost nil (0.05) between these closely related species (**Table 3**).

In Malawi, Spotted Eagle Owl had two peak months, March (nearly 60% of all calls) and November (32%). These months are also outside the given breeding season limits, August–October, in Malawi [7].

### 3.2 Some interesting owl calling observations

#### *3.2.1 Barn Owl T. alba*

M: 03/11/96 own Rooster started to call 03:55 and BO vocal immediately after the Rooster stopped 04:05.

M: 12/10/97 BO very vocal 02:00–02:05 when the Rooster started to call as well.

G: 04/08/99 BO calling 05:34—still dark and the Rooster started at the same time.

G: 06/08/99 BO calling once 06:06 and again 06:07—thrushes singing already and Rooster calling, even it was still dark, but daylight was coming soon after.

#### *3.2.2 Southern White-faced Owl P. granti*

M: 12/11/96 SwfO calling continuously 8–10 s intervals 04:20–04:30 when two Pearl-spotted Owl also calling, but one nearest the SwfO started to warn whistle "*kii-kii-kii*." Last call of PSO 04:45 when started to be light and other birds singing so loudly that it was difficult to hear the owls any longer.

M: Rooster calling 14/11/96 with the SwfO one after the other 03:00–03:30 h.

M: 20/11/96 SwfO calling 00:50 onward until 01:00 with two PSOs every minute together and unnoticed from each other.

*3.2.3 Pearl-spotted Owl G. perlatum*

M: 01/11/96 continuous calling of at least 6 owls: 3 PSO and 3 SwfO together and on top of each other 21:15–21:22 Hyena yelling, and dog barking silenced the PSO's and SwfO's, but SEO calling over the dogs and hyena 21:40–21:45—maybe that large owl silenced the smaller owls (See Discussion).

M: First gun shooting from distance 22:19 and endless drumming from 2 to 3 location. Sudden silence with dogs and owls—only Mosquitos made noise!! Dogs getting mad again 22:59 as well as the African drums. First BO called 23:00 and two PSO's continued calling 23:20–23:34.

*3.2.4 Weather impact*

M: 02/11/96 no owls 19:18–20:30 because of a heavy wind rising and a new rainstorm building up.

M: 04/08/98 BO calling next to the FAO office 18:00 despite very cold (+15C). weather.

G: 14/09/98 massive storm during the night and BO calling immediately after rain stopped at 03:30 h.

G: 30/06/99 BO calling behind the house 23:50—rainy day, but in the evening the rain stopped, bats also very active-eating mangoes!

G: 11/07/99 BO calling at 22:28—rainy day, but the rain had stopped before the owl was calling.

G: 11/08/99 BO calling 22:00 behind the house after three days of heavy rains explaining why not heard in last few days.

G: 22/08/99 BO calling 22:05 after a heavy storm, although still raining slowly.

G: 07/09/99 BO calling 19:40 after heavy rain.

G: 27/12/99–02/01/2000 very few BO calls—if any—weather relatively cool ca. +20C at lowest and often heavy winds!

G: 01/11/2001 PSO calling in daylight (07:00) in Pakalinding after heavy rain.

*3.2.5 Full Moon impact*

M: BO calling 23/04/97 00:20 but total silence during the full moon 24–26/04/97. Vocal again on 27/04/97 22:15 onward when the moon was some 60%.

M: 10/11/97 Club Makokola PSO calling in moonlight 20:00–20:10—very hot +37–38°C, humid and no wind.

G: 23/01/98 PSO calling in flight at 06:00 and again 07:00 when still very dark this time of year and relatively cold (+16°C) but cloudless sky and no wind.

*3.2.6 Some remarkable sites*

We don't knows very many sites in the world where one could hear more than one owl or a maximum of two species by sitting comfortably in a balcony chair with a glass of wine.

However, in Africa, we were lucky enough to have all the rented houses in such places, especially in Malawi (five different houses in Lilongwe and two in Cape Point in The Gambia). Some wildlife lodges were such wonders both in Malawi and in The Gambia. We want to mention especially Njobvu Safari Camp in Malawi where at least ASO, PSO, PFO, and SEO should be possible although 05/08/95 only PSO was calling several times after 23:00 h; the Dream Bird Hotel in Georgetown, The Gambia, where one can hear, for instance, on 22–23/04/2003 ASO and several PSOs; Fullady Camp where one can listen to ASO, NwfO, and PFO during the same night; Janjanbureh Bird Safari Camp wherein one night 12–13/11/2000 several voices of ASO, ESO, and PSO were recorded, and another night10–11/01/2001 a record number (6) of owl species were heard as follows: ASO, BO, MEO, NwfO, PFO, and PSO. To add a few favorite sites: Sindola Camp particularly good to hear PSOs; Tendaba Camp where ASO, BO, NwfO, PFO, and PSO are regular daily and nighttime guests, and the last but not the least Kiang West National Park that can offer ASO, GEO, MEO, and NwfO (John Clayton, *in litt.*).

## 3.3 Additional remarks on less-studied species

### *3.3.1 African Barred Owlet Taenioglaux capense*

One house in Lilongwe, Malawi, had at least five owl species calling often in the evenings. For instance, on 05/11/95 ABO calling together with the SwfO between 19.00 and 19:20 h in the full moon. The same species heard following nights but not so actively as during the full moon.

### *3.3.2 African Scops Owl O. senegalensis*

In Namibia near Windhoek, ASO calling activity was studied near the nest and no calling took place before the last light about 25 min after sunset. The male called most actively during the first hour of darkness (19:00–20:00) but some calls were heard throughout the night until 06:00 [8]. This coincides well with a Malawi record when the ASO was vocal 05/11/95 between 19:00 and 19:20 h. In The Gambia, ASO calling 19:30–20:30 near Kanifing on 15/11/2000 together with more than two NwfO's; in the Kiang West National Park ASO frequents near the camp (John Clayton, *in litt.*); near the Fullady Camp, 15/02/99 ASO calling in the evening. Same in the Tunku Creek, near Tendaba Camp, after 17:30 on 11/03/99 (Wandi Touray, *in litt.*). Janjanbureh Safari Camp 11/01/2001 AOS vocal between 03:20–04:00 but only when BO was not calling. BO vocal at 21:13; 03:20, regular 04:30–05:05 and last morning call 05:45; See Discussion); Eddy's Hotel in Farafenni 30/10/2002 ASO calling around 05:00 in the morning regularly but not so actively as nearby PSO and NwfOs; Georgetown Dream Bird Hotel 23/04/2003 AOS calling from distance between 02:00 and 03:00 h.

### *3.3.3 African Wood Owl S. woodfordii*

In a study in Kibale National Park, Uganda, it was noted that the vocalizations of AWO were more numerous during the full moon and on clear nights [9]. Very few call records were made in this study, although one AWO was recuperated at home in 1982 in Ivory Coast. In Malawi, in the hill forest of Zomba AWO was said to be very common (John Wilson, *in litt.*). In Lilongwe one AWO was calling in the dawn at 18:30 on 14/11/93. In the Kamuzu One, dam AWO was seen just before lunchtime on 26/06/95. One pellet was found under the roosting place, and it contained one house mouse

*Mus musculus*. In The Gambia, AWO was never hear but one was seen near the Abuko Education Centre 15/11/2000 (Solomon Jallow, *in litt.*).

### *3.3.4 Eurasian Scops Owl O. scops*

On MacCarthy Island in The Gambia "frog-like" calling of ESO was heard 12/11/2000. That owl calling 8 to 9 s intervals almost all night. As the owl was turning its head and changing the site, one felt there being several of them calling. Next night in the Bird Safari camp ESO was vocal 01:00–01.30 h with a very short "*grrr*" notes and another owl was answering always with whistling type of voice. Then, 02.40 ESO gave a Little Owl like "*kuiv-kuiv-kuiv-kuiv*" call and another owl responded with a similar call.

### *3.3.5 Giant or Milky Eagle Owl B. lacteus*

In Malawi, we got a young MEO what Dr. Lawrence had confiscated from beach-boys in Senga Bay on 11/09/1994 (**Figure 1**). It was estimated that the owl was born ca. 22 August. A month later the weight was already about 500 g and wing feathers showing already. Two-month old was flying fluently. This owl was in the house until 14/04/95 but stayed in the garden even after that. As a farewell show, it started calling in a tree where we had a nest box. Between 22:00 and 23:30, it was calling loudly without any breaks. That made the dogs crazy, but they calmed down when told that it is our owl. MEO kept calling even if we walked with the dogs in the garden. In July 1995, John Alder found an MEO nest in Lilongwe Nature Sanctuary, not far from the house. It was tempting to think it was our owl. It was possible only to post confirm the breeding by finding below the nest a fresh MEO wing feather on 15/11/96.

In the Gambia, MEO was seen in both Abuko and Tendaba in December 2003, but no calls were recorded; in Kiang West National Park MEO pair duetting near the camp in 1998 but timing was not taken down (John Clayton, *in litt.*); Janjanpureh Safari Camp MEO calling bouts heard two times during the night on 11/01/2002.

**Figure 1.**
*Young Milky Eagle Owl* Bubo lacteus *(left) and Spotted Eagle Owl* Bubo africanus *in their large open-air closure in our house in Malawi. Photo: Heimo Mikkola.*

### *3.3.6 Greyish Eagle Owl B. cinerascens*

After having kept three GEOs at home both in Ivory Coast (1983–1984) and one in The Gambia (2001–2002) to recover from serious mal-handling of humans, we never heard them calling. Based on that, it was concluded that GEO is a much more silent species than the closely related Spotted-Eagle Owl many of which had also been recuperated at our home in Malawi [10]. Luckily the modern sound recordings from The Gambia of breeding and non-breeding GEOs have proven that this owl is not mute. Clive Barlow's recordings will be part of an extensive study to be undertaken by Magnus Robb of the Sound recordings [11].

### *3.3.7 Pel's Fishing Owl Bubo peli*

M: One was supposed to hear PFO in Njobvu Safari Camp but on 05/08/95 this failed as it did also later in Malawi. In the Liwonde National Park, the African Cuckoo Hawk *Aviceda cuculoides* was seen to attack the PFO in September 1996 (von Bechtolsheim, *in litt.*).

G: Bao Bolon swamp 05/02/98 PFO calling at 08:00 until 09:10 h; faint but a record shot of moans (Clive R. Barlow *in litt.*).

G: Fullady Camp 15/02/99 PFO calling between 02:00–03:00 during the night.

G: Bao Bolon 11/03/99 PFO seen at 17:30 h. (Wandi Touray, *in litt.*)

G: Bao Bolon swamp 28/08/99 PFO near the broad river 10:30 h sitting in a large tree but escaping after an intensive binocular session.

G: Bao Bolon 31/10/2000 two PFOs (a pair) sitting in a tree at 09:00 drying themselves in the morning sun.

G: Kissi end of Bolon 30/09/2001 a new site for the PFO (Wandi Touray, *in litt.*)

G: Janjanpureh Safari Cam 11/01/2002 PFO calling three times during the night.

### *3.3.8 Nightlife remarks*

#### *3.3.8.1 Human activities*

M: Heavy shooting normally with AK47Kalasnikov's automatic rifle: 01/11/96 first rifle shooting from distance at 22:19 and endless drumming from 2 to 3 location; 02/11/96 shooting at 18:48; 08/11/1996 rifle shooting 22:59; 09/11/96 00:05 and 23:01; 13/11/96 at 22:08. Malawi had serious security problems due to political turmoil before the death of the old President.

M: African drumming took place often: 01/11/96 drumming started 22:03 and Spotted Eagle Owl vocal again 22:05.

In G, there was not much shooting, but Muslim prayers sometimes stopped owls calling: Two Pearl-spotted Owls calling at 02:00 at regular intervals until the Muslim Praying started in the Mosque at 05:00 when it was still dark but hot and calm on 23/04/2003.

#### *3.3.8.2 Mosquito*

Mosquitoes are scaring people in Africa as they carry malaria which has killed more people than any other disease this far. Mosquitoes make a whining sound instead of a buzz. They are most active from dusk to dawn.

M: 02/11/96 Mosquitos started whining at 21:34 when the wind was over and no rain—before that a heavy wind and a rainstorm, temperature + 22C.

M: 09/11/96 mosquito whining started already at18:46 h.

#### *3.3.8.3 Termites*

M: Incredible termite attack stopped all hearings—hundreds of female termites decided to enter our house from 23:40 onward on 01/11/96 when the temperature was +22 C.

#### *3.3.8.4 Hyena*

M: One of the most disturbing animals in owl hearing was nearby moving Spotted Hyena *Crocuta crocuta* also known as "Laughing hyena" or "Tigerwolf." Its loud "*who-oop*" call, along with maniacal laughter is among the most recognizable sounds in Africa. Whoop sound can be heard more than 5 km away. Our dogs (**Figures 2** and **3**) heard it always well before us and their crazy barking was stopping us the hearing neither the hyena nor the owls. In Malawi, hyena populations occurred in those times at reasonable densities but have since gone down due to growth in human population, habitat destruction, and reduction in prey [12].

M: Hyena calling 01/11/1996 and heavy dog barking spoiled owl listening at 20:30–20:40, "Turo" very excited and ready to go for the hyena 21:30; our dog's hyena barking started 19: 32 on 09/11/96—hyena continued at 23:10 but the dogs were too tired to bark.

M: Hyena howling agitated the dogs: 14/11/96 at 20:17 h.

M: Hyena hysteria again: 16/11/96 at 23:33 h.

#### *3.3.8.5 Spitting Cobra*

M: Somewhat disturbing was when a large snake, most likely spitting cobra *Naja* spp. took all attention for 20 min followed by the African drums at 19:29 on 13/11/1996.

**Figure 2.**
*Turo would like to taste the day-old chicken menu of the Spotted Eagle Owl. Photo: Anita Mikkola.*

**Figure 3.**
*Pepe testing the friendliness of the young Spotted Eagle Owl* Bubo africanus *in Malawi. Photo: Anita Mikkola.*

## 4. Discussion

Before commenting 2062+ owl calls recorded in Malawi and The Gambia below is short summary for each species on what the handbooks [7, 13–17] and papers [18–23] say about the calling and main breeding times of these species:

*African Barred Owlet*: Late afternoon one can hear a repetitive, mournful "*krroo, krrooo, krrooo*" call rather like that of a Ring-necked Turtle Dove *Streptopelia capicola* (Previous Cape Turtle Dove). It lacks the vivacity of the Pearl-spotted Owl's crescendo whistle [13]. Another book says: "Its call is a series of 6–10 low, whistled notes with a half-second interval between notes: '*hue-hue-hue*'; usually 2–6 second pauses between series. Series may rise and fall in volume and sometimes one series follows immediately by another. Calls mainly at night but sometimes in daylight" [7]. A third book describes the voice: "Repeated fairly high-pitched series of notes '*purr purr piu piu piu piu*', rising slightly in volume; also 2-syllabled slightly trilled '*prr-purr, prr-purr*', second note slightly higher than first." [14]. In Malawi, the breeding season is October [7] and Southern Africa from September to October [14].

*African Scops Owl*: This owl calls regularly in the evening at dusk, the female and male answering each other with an insect-like ventriloquial "*prrrup*" at approximately five-second intervals. It is a call that carries over a considerable distance and, if one is sitting by a campfire, it immediately enhances the whole bushveld atmosphere. Often, however, one may not recognize it as an owl at all [13]. The call resembles insect voices so much that most likely it was missed often on noisy African nights, especially in Malawi. In The Gambia, breeding in September [7] and Southern Africa from September to November, but also in June in Zimbabwe [14].

*African Wood Owl*: Songs almost every night and loud calls are audible over at least 500 m and show sufficient individual variation for an observer to discriminate between some individuals [19]. The male song is typically described as a rhythmic "chuckle" sequence of clear hoots, "*hoo-hoo,hu,hu, hu,hu -hu,*" the last five syllables delivered unevenly with a syncopated rhythm. Female has a higher-pitched "*eeyow*" to which the male answers by a low gruff "*hoo* or *woo*" depending on the listener's interpretation [20]. The breeding season in Sierra Leone is February [7] and in Southern Africa August to November but in Zimbabwe also one April record [14].

*Barn Owl*: In South Africa, the calling in the garden intensified during February and the beginning of March. The call most often heard is a drawn-out tremulous screech—*schrreeee*—an eerie sound which serves a variety of functions: for territorial advertisement, courtship, and a contact call [13]. The male is said to screech more often at beginning of the breeding season when courtship chases are common [7]. In The Gambia breeding September–April [7] and Southern Africa from February to May, but breeding is possible in all months [14].

*Eurasian Scops Owl*: Palearctic migrants wintering in Africa call very little during their stay making them difficult to study [15]. However, a soft frog-like croaking "*drrrr...drr..*" was heard in The Gambia and first labeled as owl x. Other owl x voice was a Little Owl *Athene noctua* like "*kiev, kiev, kiev*" notes. Only in Hungary we realized that the Gambian owl x was *O. scops* [21]. During the winter survey in Italy, it was found that the few Eurasian Scops Owls wintering there responded more strongly to playback of the Little Owl than they did to Eurasian Scops [22].

*Giant or Milky Eagle Owl*: The normal call is a very deep double hoot "*oop-poop*," almost as deep as the call of the Ground Hornbill *Bucorvus leadbeateri*. These two species could be confused, as both may roost and call from the riverine Acacia Forest at dawn and dusk. Luckily, the owl has a variety of other calls. The presumed male emits a series of short, grunting hoots "*uh.uhu-uh-uh*" and is answered by a deeper "*uh-uh*" from his mate. This duet once continued for 15 minutes [23]. Distinctive gruff hooting call is described also as "*hook-hook*"; uttered with inflated throat at variable rate and volume but in series of 1–5 hoots with an interval between series [7]. The call is said to carry for up to far as 5 km, but this is hard to believe [16]. In The Gambia breeding November–February [7] and Southern Africa from March to September, mainly June–August [14].

*Greyish Eagle Owl*: The song of male Greyish Eagle Owl recorded in Mali and Ivory Coast consists of two clear syllables and has been rendered as "*kuo-wooh*" [16] or "*koo-whoo*" [17]. The first syllable is rather explosive, and the second syllable is somewhat downward inflected, lower-pitched, and extended. This call is uttered in intervals of several seconds [13], and it is not like that of the Spotted Eagle Owl [17]. Breeding mostly from November to April almost throughout the species range [17], but in The Gambia, two well-studied nests had eggs in March [11].

*Northern White-faced Owl*: A disyllabic call is mellow fluting "*po-proo*" at 4–8 s intervals [7]. The first note is a very short, longer second note following 0.6 s [16]. In other words that voice has been described livelier as "*cuk-coooo*": a brief note, followed less than a second by another note, somewhat louder, elongated and descending. This motif sounds rather pigeon-like and is repeated at intervals of 5–12 (average 6.49) s [17]. The main call is quite different from the Southern White-faced Owl. In The Gambia, breeding takes place in February–April, but there are records also from the October–December period [7].

*Pearl-spotted Owl*: The sheer volume of its whistling call "*tiu, tiu, tiu, tiu*" is amazing for such a small bird. Notes began softly but increased gradually in intensity before achieving a penetrating crescendo. Sometimes a second owl would join in antiphonally. In addition, "*too-woop* and *tee-weep*" calls, the latter higher-pitched call that of the female; these soft calls are used by the pair to maintain contact [13]. In The Gambia breeding February–April and Malawi August–September [7].

*Pel's Fishing Owl*: Its main call is a deep sonorous hooting preceded or followed by low grunt "*hooommmmm-hut*"; repeated horn-like "*hoom-hoom*"; resonance from inflated air sacks. Sometimes male and female call in duet, male starting with grunting "*uh-uh-uhu*" building up to high "*hoommm*"; female answering by deeper hoot [17]. The hoot carries over a great distance, up to three kilometers on a calm night. Unlike many other owls, they do not become vocal at dusk and call mainly from midnight to dawn and the hoot is used throughout the year as a contact call [24]. In Nigeria, a young in a nest in February [7], in Zimbabwe breeding April, May, and October and in Botswana mainly February–April [14].

*Southern White-faced Owl*: A nuptial display consists of the bubbling hoot, and the male may approach the female along a branch while bobbing his head up and down and hooting [13]. A pair defends territory by calling; male calls regularly at dusk and dawn but also the night; female may join in. The bubbling polysyllabic "*popopopo-popeeu*" has a very fast stutter at first and the second part is more mellow, fluting, and rising in pitch. In other words, the same voice has been described as a rapid series of 5–11 (average 9) hoots, the last one being somewhat higher and accented, repeated at intervals of 7–15 (average 9) s. This song may be written "*w-h-h-h-h-h-h-oo*," pronounced as rapidly as a man may do [18]. It is quite different from the Northern White-faced Owl [7]. In Southern Africa breeding May to November, in Malawi peak is in August [14].

*Spotted Eagle Owl*: Increased hooting during the courtship period, the male's "*hoo-hoo*" being answered by the female's triple hoot "*hoo-hoohoo*," the middle "*hoo*" higher, so that the call has a pleasing cadence. Usually, the pair would duet, the female answering her mate immediately so that it sounds like a single owl hooting [13]. Both sexes may call at any time, the male usually around dusk and dawn and female in the early hours of the night [7]. In Malawi breeding season August–October [7].

From the studied 12 species, the far most vocal were Barn Owl and the Pearl-spotted Owl. The Southern White-faced Owl can also be audible most of the year, and in the captivity in Mozambique 1992, it was the most vocal of any of those owl species we have recuperated at home. Also, the less studied owl species can be more vocal than anticipated but this material is not enough to prove that.

The similarity index was low when comparing the calling hours and calling months between The Gambia and Malavi, and there was no similarity at all between two closely related White-faced Owls in The Gambia and Malawi. This may support further the separation of these owls.

Several factors are known to influence the patterns of vocal activity of nocturnal birds. One of the most obvious is the time of year, with the calling rate varying within the breeding cycle [25]. Owls are normally calling most actively especially just before breeding [4], but in this material, the peak months coincided very little with the given breeding times in Malawi and The Gambia. It is possible that the validity of the breeding times is not enough, but it is also clear that little seasonality was noted in the calling activity of the most owls studied throughout the year. This is something particular to the tropical weather conditions. The territorial calling of the Northern owls is more limited to the breeding cycle, like that of the Eurasian Pygmy Owl *Glaucidium passerinum* from March to May [1].

The overall activity pattern governs the calling as well, that is, strictly night-active species are mainly vocal in the dark while at least partly day-active species can call all around the clock. Well studied Malawi Pearl-spotted Owl is a good example of that as there are only two hours in 24 h, without any recorded calls. Month-wise the Barn Owl call all year round but has a clear break in the calling during the daytime. Similarly, White-faced Owls and Spotted Eagle Owls were not heard during the daytime.

This chapter wanted to give particulars species by species also commenting how well limited calling activity surveys are likely to reveal the actual number of existing owls.

Differences in calling rates among owls suggest that not all owls will be equally detectable using calling surveys [cf. 26]. It seems that in Africa, the Pearl-spotted Owl and Southern White-faced Owl populations and distribution should be possible to study by using the vocal surveys. They both are laud and calling actively in the evening hours. The nocturnal calling survey is not as good to map the distribution of the Barn Owls in the area as they seem to call erratically and between long intervals. Malawian data for Spotted Eagle Owl are limited but give the impression that vocal studies can serve to map the population size and distribution as the call takes place at least during the breeding times. In Kruger National Park, South Africa, it has been determined that individual African Wood Owls can be identified reliably by their vocalizations [27]. Identification of individuals by their calls has the potential for censusing, long-term population monitoring and is a valuable aid for planning the conservation of this species in Africa [27].

The influence of the weather was also studied. Heavy rain and wind are silencing the owls or at least make it impossible to hear their voices due to the background noise. Barn Owl was often calling immediately when the heavy rain and windy storm stopped. There are some examples that the temperature is not so important if the other conditions are suitable for calling, Barn Owl has been heard in +36°C as well as in +15°C; similarly, Pearl-spotted Owl records cover a similar temperature range from +37°C to +16°C.

The effect of moon luminosity on owls was also studied but with somewhat contradictory results. It seems that some owl species may increase vocal displays during full moonlight (such as the African Barred Owlet, Pearl-spotted Owl, and Southern White-faced Owl in this material) but others call less or not at all during the full moon (Barn Owl in this material). The impact of the full moon was not that obvious as the bright sky can also activate the Pearl-spotted Owl. Barn Owl started to call actively again when the moon was diminishing to 60% of its full size and its luminosity. It has been noted with other owls that they call more in the last quarter and the new moon phase of the lunar cycle [26]. In the classic Tawny Owl *Strix aluco* study in Denmark the owl called less when the moon was up than when the night was cloudy and overcast [28]. And recently studied Long-eared Owl *Asio otus* in Russia was calling both during the rising and waning phases of the moon but again no calls were recorded during the full moon [29]. It has been suggested that small mammals and even some small birds are more active on moonlight nights, with the result that owls then hunt more and call less [16].

The pitch at which an owl calls is related roughly to its body size. Small owls usually utter higher calls than their larger relatives, but sufficient for the smaller area of the territory they defend and the shorter distances over which they must communicate. Furthermore, in owls, as in other predatory birds, females are larger than males and so their calls are usually, but not always, pitched slightly lower than those of their mates (e.g., Milky Eagle Owl in this material).

Owls seem to call more frequently on still nights when there is little interference with sound transmission. The larger owls with deep voices are especially wont to call in the still hours before dawn. They may be taking advantage of layers of air of different densities—the cool dense air of the pre-dawn chill has warmer air above—that bounce back some sound from their interface and enable calls to carry over greater distances. This also ensures that they have their say before being drowned out by the dawn chorus of diurnal birds [4].

Notably, owls usually fly to prominent perches (like our garden Spotted Eagle Owl in Malawi came often on the rooftop) before they call, in this way avoiding the absorption of sound by the ground. This could entail a risk for smaller owls, whose calls might attract larger species to prey on them, but the advantage of successful communication would have to be weighed against this threat. Such interactions may explain the choice of some of the sites from which owls call, as well as the ventriloqual nature of some of their calls [4].

The predation risk, indeed, is among the most principal factors that will influence the patterns of vocal activity in owls [30, 31]. In this study, it was noted that the obvious predation risk and interference competition was altering the vocal activity of the African Scops Owl which stopped calling when the Barn Owl was active. A larger Barn Owl can be a predator that eats the smaller African Scops Owl [31]. In Malawi, it was also suspected that Spotted Eagle Owl calling silenced the Pearl-spotted and Southern White-faced Owls. And there are clear indications that Spotted Eagle Owl could prey on these smaller owls [13].

## Acknowledgements

In Malawi John Alder, Matthias Frhr. Von Bechtolsheim, and John Wilson kindly gave us some owl observations, and Clive Richard Barlow, John Clayton, and Solomon Jallow did the same in The Gambia. Wandi Touray was showing us several times Bao Bolon Pel's Fishing Owls in The Gambia and as always Alan Sieradzki sent us some old papers, we could not find in our forest library. An anonymous referee made constructive comments at the review stage which helped to improve the manuscript. We thank them all very warmly.

## Author details

Heimo Mikkola[1]* and Anita Mikkola[2]

1 University of Eastern Finland, Kuopio, Finland

2 Master of Science and Master of Arts from The University of Oulu, Finland

*Address all correspondence to: heimomikkola@yahoo.co.uk

## References

[1] Mikkola H. Milloin ja millä säällä pöllöt huutelevat (Time and weather related with owl calling). Kainuun Linnut. 1970;**1**:52-55

[2] Mikkola H. Owl. In: Campdell B, Lack E, editors. A Dictionary of Birds. Calton: British Ornithologists' Union; 1985. pp. 419-421

[3] Duncan JR. Owls of the World. Their Lives, Behaviour and Survival. Buffalo: Firefly Books; 2003

[4] Kemp A. Owls of Southern Africa. Cape Town, South Africa: Struik Winchester; 1987

[5] Gunawan KW, Hidayat AA, Cenggoro TW, Pardamean B. A transfer learning strategy for owl sound classification by using image classification model with audio spectrogram. International Journal on Electrical Engineering and Informatics. 2021;**13**(3):546-553. DOI: 10.15676/ijeei.2021.13.3.3

[6] MacNaughton SJ, Wolf LL. General Ecology. New York: Holt, Rinehart and Winston; 1973

[7] Hilary Fry C, Keith S. Urban EK. The Birds of Africa. London: Academic Press; 1988

[8] Brown CJ, Riekert BR, Morsbach RJ. The breeding biology of the African Scops Owl. Ostrich. 1987;**58**:58-64

[9] Seavy NE. Environmental correlates of African Wood-owl calling activity in Kibale National Park. Uganda. Journal of Raptor Research. 2004;**38**(3):208-213

[10] Mikkola H. Owl knowledge and Beliefs in Africa. Tyto. 2021;**26**:9-36

[11] Barlow CR, Mikkola H, Wink M, Brohaugh E, Brohaugh A. Molecular evidence for the taxonomic status of the Greyish Eagle Owl Bubo cinerascens and a breeding study in The Gambia. Malimbus. 2022;**44**(1):19-29

[12] Mills G, Hofer H. Hyenas: Status Survey and Conservation Action Plan. IUCN/SSC Hyena Specialist Group. Gland, Switzerland 1998 (ISBN 978-2-8317-0442-5)

[13] Steyn P. A Delight of Owls—African Owls Observed. Dover, New Hampshire: Tanager Books; 1984

[14] Maclean GL. Roberts' Birds of Southern Africa. Cape Town: The Trustees of the John Voelcker Bird Book Fund; 1985. p. 848

[15] Robb M. Undiscovered Owls. A Sound Approach Guide. 308 pages. 2015. The Sound Approach, Poole Dorset. (ISBN: 978-90-810933-7-8).

[16] Mikkola H. Owls of the World Enhanced e-book. 528 pages. 2014. Bloomsbury/Christopher Helm London. (ISBN: 978-1-4729-0592-5).

[17] König C, Weick F & Becking J-H. Owls of the World. Second Edition. 2008. Christopher Helm, London.

[18] Van der Weyden WJ. Geographical variation in the territorial song of the White-faced Scops Owl *Otus leucotis*. Ibis. 1973;**115**:129-131

[19] Delport W, Kemp AC, Ferguson WH. Vocal identification of individual African Wood Owls *Strix woodfordii*: a technique to monitor long-term adult turnover and residency. Ibis. 2002;**144**:30-39

[20] Steyn P, Scott J. Notes on the breeding biology of the Wood Owl. Ostrich. 1972;**44**:118-125

[21] Mikkola A, Mikkola H. Voice and daytime calling of Scops Owls (*Otus scops*). Ornis Hungarica. 2015;**23**(2):52-55. DOI: 10.1515/orhu-2015-0014

[22] Mori E, Menchetti M, Ferretti F. Seasonal and environmental influences on the calling behaviour of Eurasian Scops Owls. Bird Study. 2014;**61**:277-281

[23] Brown LH. Observations on Verreaux's Eagle Owl *Bubo lacteus* (Temminck) in Kenya. Journal of East African Natural History Society. 1965;**25**(2):101-107

[24] Brown LH. Observations on Pel's Fishing Owl *Scotopelia peli*. Vol. 96. UK: Bulletin of British Ornithological Club; 1976. pp. 49-53

[25] Lourenço R, Goytre F, Delgado MM, Thornton M, Rabaça JE, Penteriani V. Tawny owl vocal activity is constrained by predation risk. Journal of Avian Biology. 2013;**44**:001-008. DOI: 10.111/j.1600-048X.2013.00157.x

[26] Ganey JL. Calling behaviour of Spotted Owls in Northern Arizona. The Condor. 1990;**92**:485-490

[27] Kemp AC, Kemp MI. The use of sonograms to estimate density and turnover of Wood Owls in riparian forest. Ostrich Supplement. 1989;**14**:105-110

[28] Hansen L. Natuglens (*Strix a. aluco* L.) døgn- og årsrytme. Dansk Ornithologisk Forenings Tidsskrift. 1952;**46**:158-172

[29] Andreychev A, Lapshin A, Kuznetsov V. Vocalization of the Long-eared owl *Asio otus* (Strigiformes, Strigidae) in the Middle Volga, Russia. Biodiversitas. 2021;**22**(12):5325-5330. DOI: 10.13057/biodiv/d221213

[30] Franchuk MV, Yanenko VO. The abundance and spatial distribution of the Eurasian Pygmy Owl, *Glaucidium passerinum* (Strigiformes, Strigidae), in Rivnenskyi Nature Reserve, Ukraine. Proceedings of the Zoological Museum, Kyiv. 2018;**49**:16-23

[31] Garcia D, Trujillo D, Parpal L. Primeros casos de depredación de Lechuza común *Tyto alba* y Búho chico *Asio otus* sobre Autillo Europeo *Otus scops* (Baleares-España). Anuari Ornitològic de les Balears. 2007;**22**:97-101

Section 4

# Country Biodiversity, Population Changes and Conservation Studies

Chapter 4

# Qualitative and Quantitative Changes in a Guild of Forest Owls: Eurasian Pygmy Owl (*Glaucidium passerinum*), Ural Owl (*Strix uralensis*), Tawny Owl (*Strix aluco*), Boreal Owl (*Aegolius funereus*) at Kamenný Hrb – Bankov Site in Volovské Mountains Near Košice Town, Eastern Slovakia, between Years 1989 and 2021

*Samuel Pačenovský and Alexander Kürthy*


## Abstract

A guild of forest owls was evaluated in 1989 – 2021 in a 5.2 km$^2$ site in Slovakian Volovské mountains in Western Carpathians. Only the Eurasian Pygmy Owl (*Glaucidium passerinum*) declined in this near 30 year period from 8 to 5 territories and that local decline is referred to the increased presence of Tawny Owl (*Strix aluco*). Clear decline of calling activity of Eurasian Pygmy Owl in sympatric areas with Tawny Owls was also noted. Tawny Owl almost doubled its population from 3 to 5 territories occupied in 2017–2021 and the Boreal Owl (*Aegolius funereus*) was almost absent during 1989–1994 but occupied 8 breeding territories in years 2017–2021. Tawny Owl has a tendency of spreading to higher elevations, while Boreal Owl has an opposite tendency. Boreal Owl seeks suitable breeding habitats in old fir-beech forests with fir and oak stands and Black Woodpecker (*Dryocopus martius*) holes in old beeches as low as below 500 m a.s.l. Due to good populations of small mammals, Ural Owl (*Strix uralensis*) has been able to maintain stable populations with ca. 8 pairs in the study area between years 1989–2021. Good food situation also attracted some Boreal Owls to lowest known elevation limit of the species in Slovakia. Further research is needed, on a larger scale, to support the population trends documented in this paper.




IntechOpen

**Keywords:** Eurasian pygmy owl, boreal owl, tawny owl, Ural owl, densities, trends, Western Carpathians, Slovakia

## 1. Introduction

The site Kamenný hrb – Bankov is situated near the town Košice, in Volovské mountains, in Eastern Slovakia, in elevation 420–550 m. Volovské mountain range belongs to W Carpathians. The study site can be characterised as a hilly, forested area, overgrown with most forests over 80 years of age (nowadays already over 100 years old). The size study area is approx. 5.2 $km^2$. Composition of forests is formed by naturally grown oak (*Quercus petraea*), horbeam (*Carpinus betulus*), beech (*Fagus silvatica*), fir (*Abies alba*) and lime (*Tilia crdata*) stands, with sycamore (*Acer pseudoplatanus*) in higher elevations and with planted stands of spruce (*Picea abies*) of the same age, as other trees. Coniferous and mixed forests (spruce and fir-beach) dominate the ridge and its northern slope and oak and horbeam forests dominate southern expositions and the lowest elevations, but even oak and fir stand to grow side-by-side naturally, and the site is one of the lowest areas with naturally growing fir in Slovakia. The forest is not homogenous, it is interrupted by smaller and middle-sized meadows, clear-cuts and forest nurseries. The forest is harvested in a moderate, sustainable way, mainly due to its proximity to the town and also due to its main function – to serve as a recreation area for people from the town, with a number of forest tracks. Guild of forest owls at the site consists of 4 species: Eurasian Pygmy Owl (*Glaucidium passerinum*), Ural Owl (*Strix uralensis*), Tawny Owl (*Strix aluco*), Boreal Owl (*Aegolius funereus*), while the last species is a newcomer at the site because in period 1989–1994, only a few data of the species were found and no territorial males. Eurasian Pygmy Owl and Boreal Owl is in Slovakia traditionally associated with coniferous and mixed forests in higher mountains from 400 m a.s.l. up to forest limit in 1500 m a.s.l. [1–5], while Ural Owl and Tawny Owl is associated mostly with mountains and beech or oak forests, rarely broad-leaved lowland floodplain forests [1, 6], even if in some areas they occupy also mixed or even coniferous stands in mountains [7]. Populations of owls are regularly monitored at the site. First ornithological data on the occurrence of Eurasian Pygmy Owl and Ural Owl in the area come from the 1970s years by Mošanský (1982) [8], Danko (1988) [9], Pačenovský (1981) [10] and of the Tawny Owl from 1980s years by Mošanský (1982) [8], Takáč (1982) [11] and occurrence of the Boreal Owl from the site was previously not known. The site is one of the lowest areas in Volovské mountains with distribution of Eurasian Pygmy Owl, where two nests situated in 520 and 470 m elevation found and checked in years 1989, 1990, 1991, 1994, 1997 and 2009 were regarded as the lowest known nest sites of the species for Slovakia [5]; as well as one of the lowest known areas with distribution of the Boreal Owl. Thirty-year-long observation of the site prepared a possibility to follow population trends of 4 owl species occupying the site and most of them have shown rather steep population changes within those periods.

## 2. Qualitative and quantitave changes in a guild of forest owls Eurasian pygmy owl, ural owl, tawny owl, boreal owl at kamenný hrb – bankov site

Population of Eurasian Pygmy Owl (thereafter only Pygmy Owl) at the site Kamenný hrb – Bankov has been continually monitored from 1989 till present time, till year 2021

[5, 6, 12–16] but the most intensively in years 1989–1991. As an example of intensity of its monitoring in that period, we can mention years 1989–1994, when the site was visited 189 times (in 1989 realised 109 site visits/99 records), with a result of 161 records of the species at the area, even if over a half of site-visits was done in order to observe activity of owls at two occupied nest-sites. Local population of the Ural Owl was during the same period also under regular control, during years 1989–1991 by both authors, and from the 1980-ies till present time by local ornithologist Jozef Mihók, who placed a number of nest-boxes in the area, to support the breeding of the species. A low number [3–5] of nest-boxes was placed in the area in year 1991 also for the Pygmy Owl and for the Boreal Owl by first author of the paper, but these nest-boxes were not occupied by these species. In Slovakia breeding of the Pygmy Owl in a nest-box is very unusual, so far only one case was found [5, 9] and all other occupied nests in the country were found in holes excavated by Great-spotted Woodpecker (Dendrocopos major) or, especially in higher elevations also by the Eurasian Three-toed Woodpecker (Picoides tridactylus) and just very exceptionally breeds also in natural cavities – just 2 known cases so far [5]. On the other hand, breeding of the Boreal Owl in nest-boxes is quite common, as well as breeding of Tawny and Ural owl [1], but at study area, Kamenný hrb-Bankov nest-boxes were utilised only by Ural Owls, besides of natural raptor nests. Breeding of two pairs of Pygmy Owl was documented in Kamenný hrb area during years 1989–1994 [5, 7, 12–14] and in one territory breeding was documented also later on, in years 1997 and 2009 [5]. Also, food consumption, breeding biology and notes from ethology were evaluated from nests found in Volovské mountains [12], as well as course of autumn mating of the species was evaluated [16]. Curiosity of these breeding records is, that 5 subsequent breeding attempts were found in the same owl territory marked „B1“: in years 1990, 1991, 1994, 1997 and 2009, while during years 1990–1997 the owls used for breeding the same oak tree, even if not in every year they utilised the same woodpecker cavity. It is almost sure, that during these years (1990–1997) more than 1 male used that territory – that presumption is likely because of short life-span of the species and it was proved, that in years 1990 and 1991 another female took part in the breeding because in both years breeding female birds were captured and ringed and the female breeding in 1991 in the same tree as in previous year had no ring. Distribution of the 8 identified breeding territories of Pygmy Owls in Kamenný hrb site during years 1989–1994 is shown at **Figure 1**. Examples of scetches describing site-visits with recorded observations of Pygmy Owls are shown at **Figures 2** and **3**. Distribution of Pygmy Owl territories at Kamenný hrb site in years 2017–2021 shows a very different picture (**Figure 4**). Despite a quite intense search after the species, especially during autumn 2020 and winter 2020–2021 only 4 occupied territories were located and effectiveness of locating of the species was very low: only 4 positive occurrences of the species despite 16 site-visits devoted to an intensive search for the species, including imitations of advertising calls. Lower density of Pygmy Owl at the site could be best explained by the already described phenomenon [17, 18] of increased competition with Tawny Owl – see **Figures 5** and **6**, that caused e.g. local extinctions of Pygmy Owl in parts of W Germany after distribution of Tawny Owl to higher elevations, what almost happened also here, at Kamenný hrb site. Very low vocal activity of the species in years 2020–2021 is probably as well a result of competition with Tawny Owl, the species remains secretive even during autumn mating and winter season, only in late autumn (XI, XII) and early spring (II) was detected some territorial activity. Only 23% of site visits were effective to locate calls of the species in years 2020–2021 in comparison to 90.8% of positive site visits in 1989 and 83% of positive visits in years 1989–1994. On the other hand, occurrence of the Tawny Owl increased at Kamenný hrb site from 4 occurrences annually (females only) in 1989 to 7–10 registrations/year (regular territorial

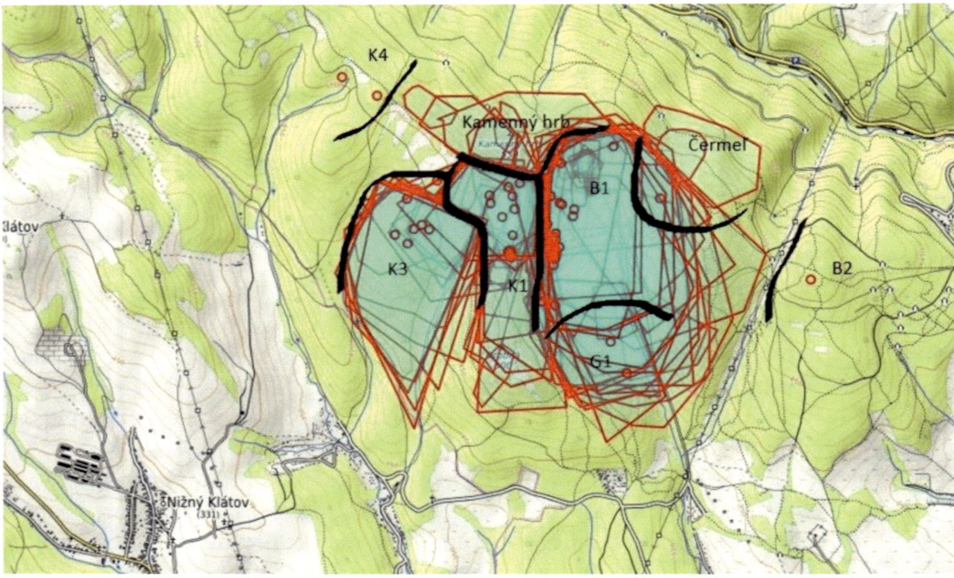


**Figure 1.**
*Distribution of 8 identified breeding territories of Eurasian pygmy owl (Glaucidium passerinum) in years 1989–1994 at the site Kamenný hrb – Bankov near Košice. Data were excerped from online databasis Aves-symfony of SOS/BirdLife Slovakia – 161 data. Names and symbols of identified territories of G. passerinum: B1, K1, K3, K4, G1, Kamenný hrb, Čermel', B2. Known limits between individual territories are marked with black solid lines. In 2 of the territories even breeding nest cavities were found: in territory K1 nest site was known in years 1989 and 1990; in territory B1 nest sites was found in years 1990, 1991, and 1994.*

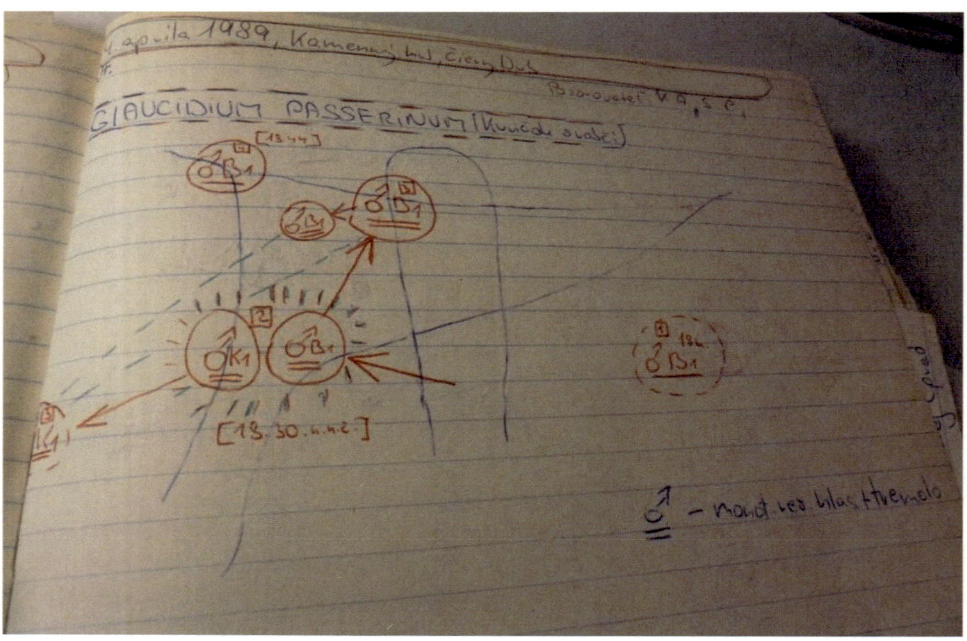


**Figure 2.**
*Example of original field note from year 1989, describing a detailed evening observation of activity of 2 calling males of Eurasian pygmy owl (B1 and K1) at border-line of the two territories (A. Kürthty).*

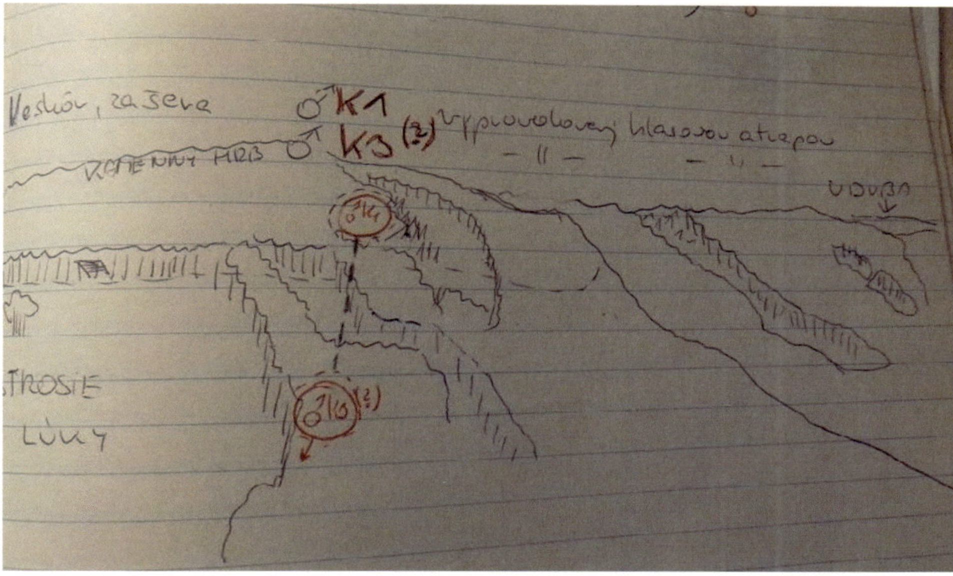


**Figure 3.**
*A "three dimensional" note of observation at a border of 2 Eurasian Pygmy Owl territories, possibly K1 and K3, in March 1989. (A. Kürthy).*

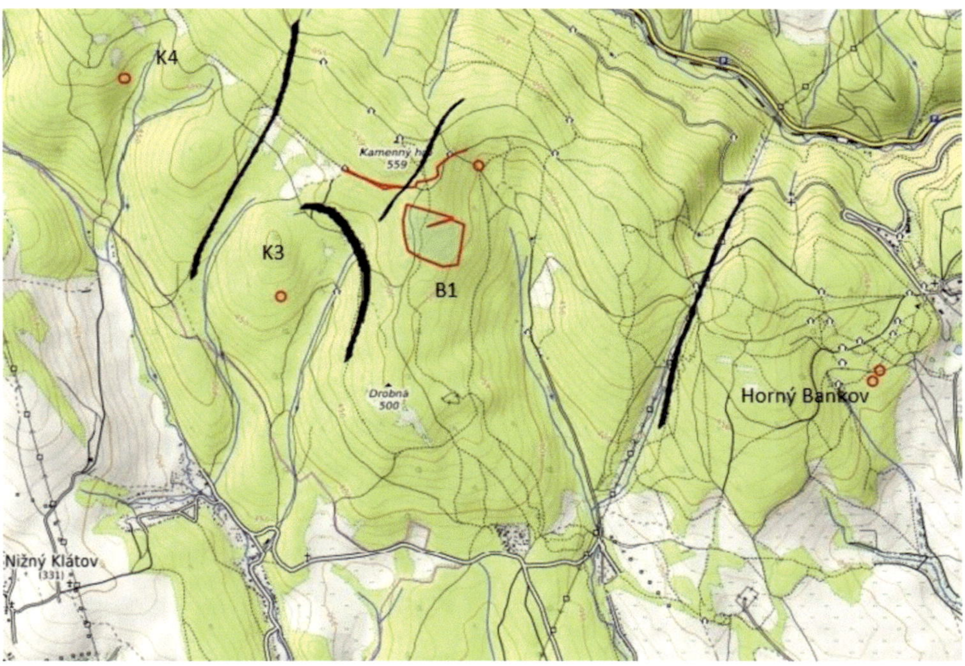


**Figure 4.**
*Distribution of 4 identified breeding territories of Eurasian pygmy owl (Glaucidium passerinum) in years 2017–2021 at the site Kamenný hrb – Bankov near Košice town. Data were excerped from online databasis Aves-symfony of SOS/BirdLife Slovakia – 8 records. Names/codes of known territories and years, when corresponding occupied territory was checked: B1 (2017, 2020, 2021), K3 (2020), K4 (2020), Horný Bankov (2019, 2020). Known, or supposed borders between individual territories are marked with black solid lines.*

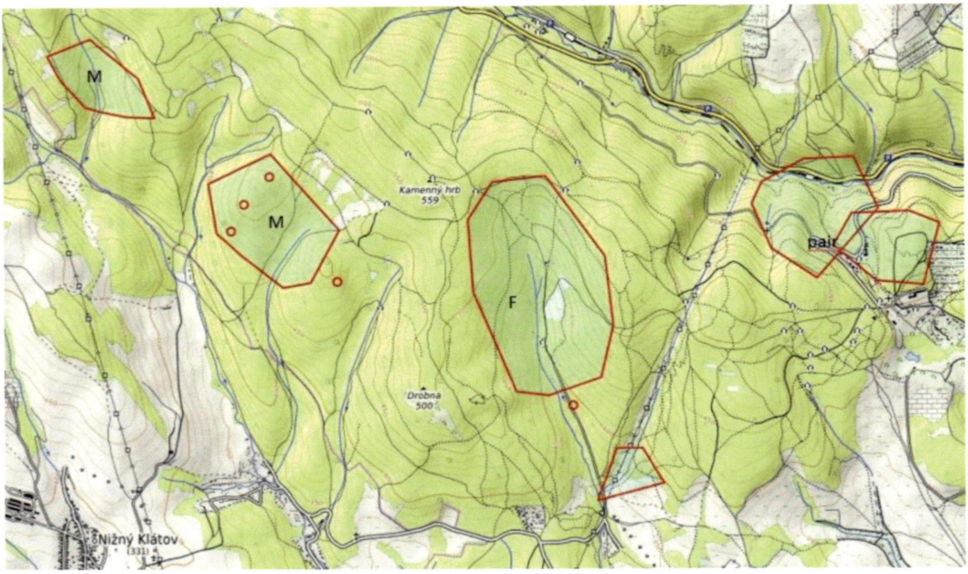


**Figure 5.**
*Known distribution of occupied territories of the tawny owl (Strix aluco) in years 1989–1994 at the site Kamenný hrb – Bankov near town Košice. Data were excerped from online databasis Aves-symfony of SOS/BirdLife Slovakia – 14 data. With symbol „M" were marked breeding territories occupied by territorial males; with symbol „F" is marked an area where only occurrence of a female bird was found; with a symbol „pair" is marked a territory occupied by a pair with detected nest site.*

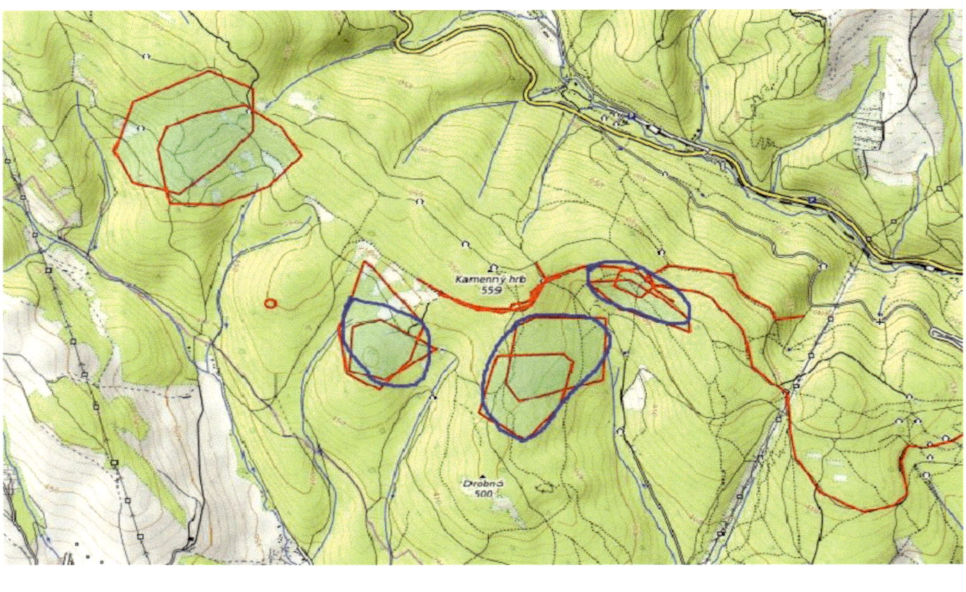


**Figure 6.**
*Known distribution of occupied territories of the tawny owl (Strix aluco) in years 2017–2021 at the site Kamenný hrb – Bankov near town Košice. Data were excerped from online databasis Aves-symfony of SOS/BirdLife Slovakia – 17 data. Number of occupied territories after 30 years moderately increased, present number of supposed breeding pairs occupying the area is 5 according to known territories defended by territorial males. Main difference is against situation from years 1989–1994 is presence of at least 2 territories of males at the main ridge to east from altitude quote "Kamenný hrb", as well as at the side-ridge Kamenný hrb – Drobná and a following hill to the W – All 3 marked with blue circles; where in the previous period 1989–1994 about 30 years earlier territorial occurrence of the species was unknown.*

males) in years 2017–2021 and that increase of territorial activity of Tawny Owls was the most intense in 5 territories of Pygmy Owls marked in 1989 as„B1", „K1", „K4", „Kamenný hrb" and„Čermel"" (**Figure 1**), where territorial presence of the Tawny Owl was in years 1989–1994 almost unknown, irregular and very rare (see **Figure 4** with distribution of Tawny Owl for years 2017–2021). Also, occurrence of the Pygmy Owl at Horný Bankov – the easternmost territory occupied in 2017–2021 (**Figure 2**) is a good example of redistribution of the species to suboptimal, highly unusual habitat for the species: 80 years old oak-horbeam forest with almost no conifers, but with a lack of Tawny Owls (but occupied by Ural Owls). Another Pygmy Owl, the westernmost at the site, at **Figure 1** shown as territory„K4" sought a„shelter" from a neighbouring Tawny Owl and has literally stitched on between 2 Ural Owl territories (see **Figures 4** and **8**), in order to get protection against its Tawny Owl neighbour, moving its territory inside two territories of Ural Owls. Even if Ural Owl is able to predate Eurasian Pygmy Owls [17] and even in one of nest sites of Pygmy Owl observed at Kamenný hrb, in territory„K1" was almost sure, that one of the 5 fledglings was predated by Ural Owl [6], that risk is apparently lower for the Pygmy Owl, as close neighbourhood of the more dangerous Tawny Owl [6], which is known to be a foraging generalist and a frequent predator of birds [19]. Even if Boreal Owl does not mean direct threat to Pygmy Owl, their main prey items can overlap (small forest mammals – mice, voles, shrews), thus steep population growth of the Boreal Owl at Kamenný hrb after the year 2000 could also possibly contribute to worsening of local conditions for Pygmy Owl. Distribution and densities of the corresponding 4 owl species: Pygmy Owl, Ural Owl, Tawny Owl and Boreal Owl at Kamenný hrb – Bankov site during last 30 years are shown at **Table 1**.

Distribution of territories, nor overall density of Ural Owls, as the most dominant of the 4 observed owl species at Kamenný hrb site did not change substantially during previous 30 years – see **Figures 7** and **8**. Most of Ural Owl territories remained at the same sites as previously and density of the species also remained the same. The Boreal Owl at Kamenný hrb site went through an expressive transition from an irregular non-breeding visitor (**Figure 9**) in years 1989–1994 to a regularly occurring population of 7–8 all-year-round present territorial males (possibly pairs) in years 2017–2021 (**Figure 10**). That change is contradictory to planet-warming process, because these owls were forced to lower elevations of the mountain range apparently from higher elevations. But if we address that phenomenon from a broader regional scale of all Volovské mountains range (140,000 ha), where the Bankov-Kamenný hrb site belongs, the explanation is simple: most of spruce forests in elevations over 1000 m in central and western part of Volovské mountains with viable populations of

| Owl species | territories 1989–1994 | density/km$^2$ 1989–1991 | No of records | territories 2017–2021 | density/km$^2$ 2017–2021 | No of records | trend |
|---|---|---|---|---|---|---|---|
| Glaucidium passerinum | 8 | 1.5 | 161 | 4 | 0.74 | 8 | −2 |
| Strix uralensis | 8 | 1.5 | 69 | 8 | 1.5 | 41 | stable |
| Strix aluco | 3 | 0.56 | 14 | 5 | 0.93 | 17 | (+2) |
| Aegolius funereus | 0 | 0 | 6 | 8 | 1.5 | 36 | (+2) |

**Table 1.**
*Number of occupied territories and densities of Ural owls, tawny owls, Eurasian pygmy owls and boreal owls at Kamenný hrb – Bankov between years 1989 and 2021.*

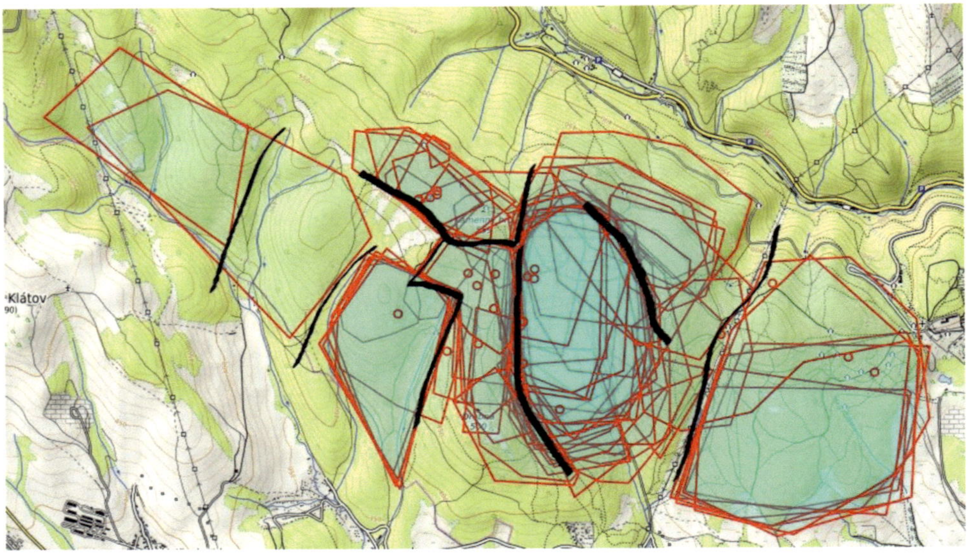


**Figure 7.**
*Known distribution of occupied territories of the Ural owl (Strix uralensis) in years 1989–1994 at the site Kamenný hrb – Bankov near town Košice. Data were excerped from online databasis Aves-symfony of SOS/ BirdLife Slovakia – 69 data. Even if specifical nest sites were not identified, it is very likely that minimally around 8 pairs has bred at the site in that period. Known or supposed borders between individual breeding territories are marked with black lines.*

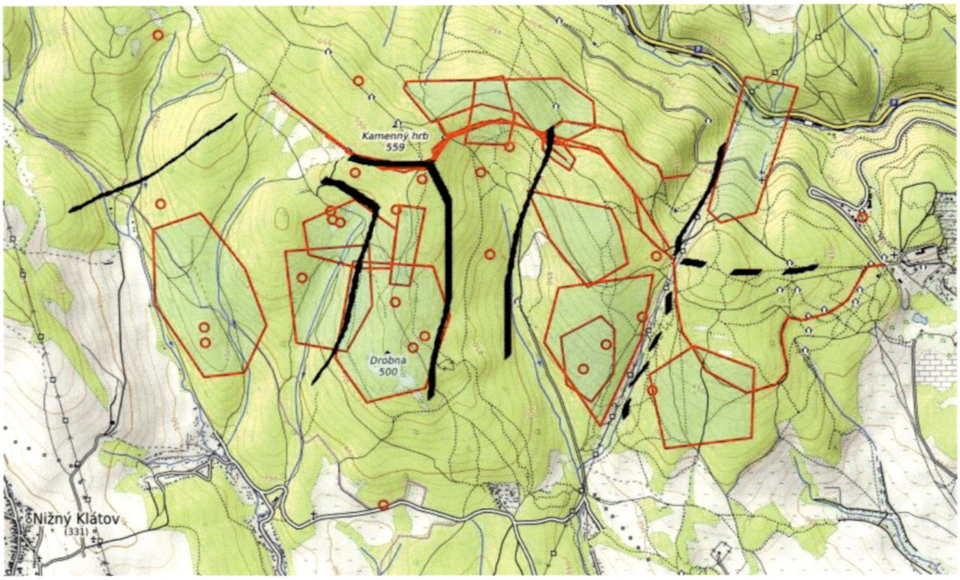


**Figure 8.**
*Distribution of occupied territories of the Ural owl (Strix uralensis) in years 2017–2021 at the site Kamenný hrb – Bankov near town Košice. 27 distribution data are displayed at detailed DFS map. The population consists of minimally 8 occupied territories of breeding pairs. In some territories even occupied nest-boxes are known. Distribution of individual territories is almost identical as in period 1989–1994.*

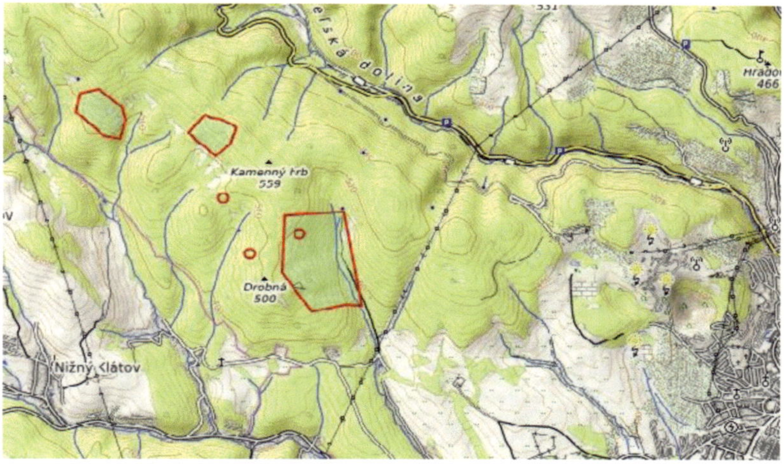


**Figure 9.**
*Sites with sporadical occurrence of the boreal owl (Aegolius funereus) at the same site Kamenný hrb – Bankov, between town Košice in the east and village Nižný Klatov in the south; during period 1989–1994. Data were excerped from online databasis Aves-symfony of SOS/BirdLife Slovakia – 6 data. None of determined birds was a territorial male.*

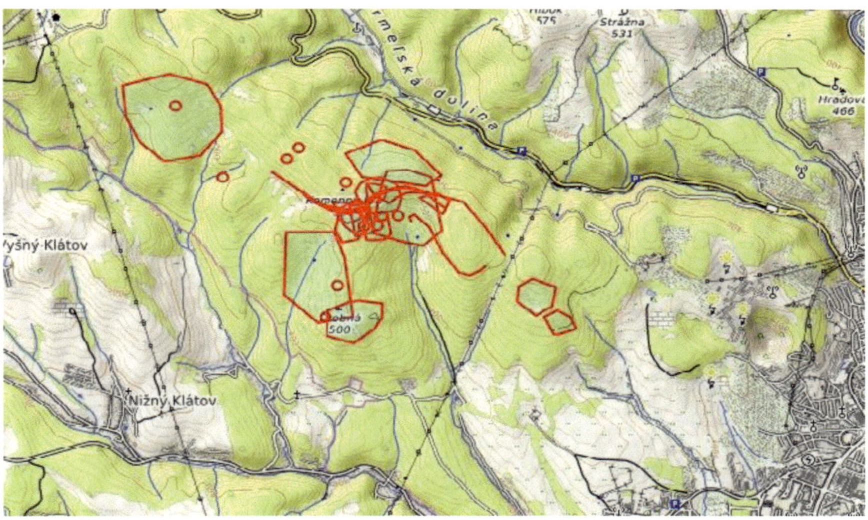


**Figure 10.**
*Sites with occurrence of the boreal owl (Aegolius funereus) at Kamenný hrb – Bankov, between town Košice in the east and village Nižný Klatov in the south; during period 2017–2021. Data were excerped from online databasis Aves-symfony of SOS/BirdLife Slovakia – 36 data. Data include 36 occurrences and approx. 8 territorial males. The species is now year-round present in the area as a territorial bird with supposed breeding, number of occurrences increased after 30 years (between 2017 and 2021 in comparison with period 1989–1994) from 6 to 36 and number of occupied territories from 0 to 8.*

Boreal Owls were in large extent destroyed and removed, between years 2000–2010 approx [14], thus occurrence of that owl species in elevations as low, as 400–600 m at Kamenný hrb site only reflects the tendency of the species looking for new habitats, instead of the destroyed habitats in higher elevations of the same mountain range. Possibly good local populations of small forest mammals, such as rodents and shrews at Kamenný hrb, able to maintain a good population of Ural Owls for the last 30 years (see **Figures 5** and **6**), as their main food source; as well as numerous Black Woodpecker (Dryocopus martius) cavities in old beech stands at Kamenný hrb site could attract Boreal Owls even to these quite unique – for the species unusually low elevations – at least from a Central-European perspective.

## 3. Broader regional ties of the observed sites to Volovské mountains

The Kamenný hrb – Bankov site belongs to a broader mountain range of Volovské mountains, of area 1240 km$^2$, reaching elevations 1322 m (**Figures 11–13**). In the whole range, Ural and Tawny Owls still survive with a stable trend, despite forest logging, but

**Figure 11.**
*Nest site of the Eurasian pygmy owl at Suchý vrch in Starovodská valley in year 2013, 830 m elevation. Photo S. Pačenovský.*

**Figure 12.**
*Breeding territory of Eurasian pygmy owl at Suchý vrch in Starovodská valley in central part of Volovské mountains, fir-beech forest in 830 m elevation, June 2013. Photo S. Pačenovský.*

**Figure 13.**
*Starovodská valley in central part of Volovské mountains, July 2013. At opposite slope over 100 years old, natural, mixed forests are visible, as well as other parts of the ridge, denudated after removal of dying spruce forests, 800–1300 m a.s.l. Photo: S. Pačenovský.*

Boreal Owls and also probably Pygmy Owls show a moderate long-term decline due to loss of spruce forests above 1000 m elevation as a consequence of climate change and bark-beatle infestations. As profound population decline of Pygmy Owl, as we observed at Kamenný hrb – Bankov site in other, higher elevations of the range was not found even during last decade (2010–2021), with exception of highest areas in central and western part of the range, where vast habitat degradation was observed in spruce forests. Other populations of all 4 owl species in fir-beech forests are still stable. Latest population estimates for whole Volovské mountains range were 130–155 pairs for the Pygmy Owl (locally −1 popul. trend), for the Ural Owl 130–190 pairs (stable popul. trend), for the Boreal Owl 130–180 pairs (−1 popul. trend) and for the Tawny Owl was not stated a precise population estimate [14].

## 4. Discussion

Food items of Pygmy Owl, collected in years 2010–2014, including three nests from Volovské mountains, from elevations 840–1040 m a.s.l. (Kojšovská hoľa, Tupý vrch, Starovodská valley) were evaluated [20]. Food items from these 3 nests contained remains of 43 specimens of 6 mammal species and 97 specimens of 22 bird species (**Annex 1**). Surroundings of the nest in Starovodská valley are shown in **Figures 11–13**. Food analysis from 2 nests of Pygmy Owl was earlier realised also exactly at the site Bankov – Kamenný hrb from years 1989–1993 [13] and food supply contained several species of small forest mammals and Passerines. These data support an assumption, that local decline of the Pygmy Owl at observed study site Bankov-Kamenný hrb was not caused by food shortage. Even habitat quality did not change considerably during that 30 year period, apart from the fact, that the forest gained 30 years of its mature age. On the other hand, good populations of small mammals in the forest and open habitats (meadows, clear-cuts), able to maintain good and stable populations of Ural Owls could possibly attract some Boreal Owls to as low areas as that site, to lowest known elevation limit of the species in Slovakia.

Density estimates of owls for Volovské mountains are actually being evaluated and an extensive monitoring scheme was realised in year 2021 to get population data for all 4 species from different sites of the range in its highest elevations over 700 m (700–1300 m a.s.l.); the results will soon be published. Could be assumed, that if Boreal Owls tend to move from highly degraded spruce forests at the ridge of Volovské mountains to lower elevations (e.g., Kamenný hrb area in 400–600 m elevation), some Pygmy Owl populations will also follow this trend. Occurrence of the species and even temporary breeding attempts in the Aggtelek Karst in Hungary with first breeding record of Pygmy Owl for Hungary [21] could be an example of that kind of process because Aggtelek karst is situated only 13 km from Southern edge of Volovské mountains and both ranges are interconnected through forested plateaus of the Slovak karst with patches of conifers (and with occurrence of Pygmy Owls in these patches of conifers).

## 5. Conclusions

Only the Pygmy Owl declined after a near 30 year period (steep decline by −2 from 8 to 5 territories); its local decline is referred to increased presence of Tawny Owl at the site. Only 23% of site visits were effective to locate calls of the species in years 2020–2021 in comparison to 90.8% of positive site visits in 1989 and 83% of positive visits in years 1989–1994. This fact was probably caused by decline of calling activity of Pygmy Owl in sympatric areas with Tawny Owls. Two other species showed steep population increase +2: the Tawny Owl almost doubled its population at the site, from 3 territories (mostly females only) in 1989–1994 to 5 territories occupied in 2017–2021 and the Boreal Owl was almost absent during 1989–1994 and appeared as regularly occurring territorial species with 8 occupied breeding territories in years 2017–2021. In case of Tawny Owl, there is a tendency of spreading to higher elevations and in the case of Boreal Owl an opposite tendency, looking for suitable breeding habitats in old fir-beech forests with fir and oak stands and Black Woodpecker holes in old beeches as low as below 500 m a.s.l., presumably due to profound habitat loss going on recently in higher, central and W parts of the same mountain range, in elevations about 1000 m a.s.l., where optimal conditions for the species were worsened after year 2000 due to large-scale drying up and destruction of spruce forests. One species, the Ural Owl remained stable at the site between years 1989–2021 with a continual total population of about 8 pairs. Wider regional comparison of the site at the scale of whole Volovské mountains range was given in chapter 3.

Earlier published studies of diet of Pygmy Owl [20] in years 2012–2015, including data from 3 nests located in higher elevations of Volovské mountains revealed an existing wide food supply for the species, including at least 6 mammal and 22 bird species found in diet of these 3 Pygmy Owl pairs. These data support an assumption, that local decline of the Pygmy Owl at observed study site Bankov-Kamenný hrb was not caused by food shortage. On the other hand, good populations of small mammals in the forest and open habitats (meadows, clear-cuts), able to maintain good and stable populations of Ural Owls could possibly attract some Boreal Owls to as low areas as that site, to lowest known elevation limit of the species in Slovakia. It has been concluded that some of quite recent, new, already published data on occurrence of Pygmy Owl in its lower distribution limit, e.g., as those on first documented breeding of the species in Aggtelek karst in N Hungary from 2011 [21] could be explained by a partial population decline and large-scale habitat loss going on in some areas of Volovské mountains, located only 13 km to N from edge of the Aggtelek karst area in Hungary. Still, the study area of 5.2 km$^2$ is quite small, so wider conclusions on population trends of these 4 owl species should be taken carefully and further research is needed, on a larger scale.

## Acknowledgements

The authors are grateful to a number of collaborators, helping with intense field monitoring of owls in years 1989–1994, namely especially to †Róbert Leščinský, Gabriel Lešinský, Róbert Olejár, Roman Rajštetter, Karol Takáč and many others, during period 2017–2021 we are grateful first of all to field collaborators Róbert Olejár, Karol Takáč and Ján Tencer.

## Annex 1

Annex 1 Food items of pygmy owl, collected in years 2010–2014, at three nests from Volovské mountains, from elevations 830–1040 m a.s.l. (Kojšovská hoľa, Tupý vrch, Starovodská valley). After: Šotnár et al. [20].

| **Samples (Prey taxon)** | **1** | **2** | **3** | **Σ** |
|---|---|---|---|---|
| Common shrew (Sorex araneus) | 1 | | | 1 |
| Eurasian pygmy shrew (Sorex minutus) | 1 | | | 1 |
| Hazel dormouse *(Muscardinus avelanarius)* | | 1 | | 1 |
| Northern birch mouse Sicista betulina | | 1 | | 1 |
| Yellow-necked mouse (Apodemus flavicollis) | 4 | 9 | 3 | 16 |
| Bank vole (Myodes glareolus) | 5 | 14 | 4 | 23 |
| **Mammals (Mammalia)** | **11** | **25** | **7** | **43** |
| White wagtail (Motacilla alba) | | 3 | | 3 |
| Dunnock (Prunella modularis) | | | 1 | 1 |
| Common whitethroat (Sylvia communis) | | 1 | | 1 |
| Blackcap (Sylvia atricapilla) | | | 2 | 2 |
| Common chiffchaff (Phylloscopus collybita) | | 3 | 1 | 4 |
| Goldcrest od Firecrest *(Regulus sp.)* | 1 | 3 | 3 | 7 |
| Common redstart (Phoenicurus phoenicurus) | | 1 | 1 | 2 |
| European robin (Erithacus rubecula) | 1 | 3 | 2 | 6 |
| Long-tailed tit (Aegithalos caudatus) | | 4 | | 4 |
| Great tit (Parus major) | | 3 | 3 | 6 |
| Coal tit (Periparus ater) | | 13 | | 13 |
| Blue tit (Cyanistes caeruleus) | 1 | 1 | 5 | 7 |
| Crested tit (Lophophanes cristatus) | | 1 | | 1 |
| Marsh tit (Poecile palustris) | | 2 | 1 | 3 |
| Willow tit (Poecile montanus) | | 1 | 4 | 5 |
| Eurasian nuthatch (Sitta europaea) | 2 | 4 | | 6 |
| Eurasian treecreeper (Certhia familiaris) | | 5 | | 5 |
| Northern wren (Troglodytes troglodytes) | | 6 | | 6 |
| Yellowhammer (Emberiza citrinella) | | 1 | 1 | 2 |
| Common chaffinch (Fringilla coelebs) | 1 | 5 | 3 | 9 |
| Common linnet (Linaria cannabina) | 1 | | | 1 |
| Eurasian bullfinch (Pyrrhula pyrrhula) | 1 | | | 1 |
| Passerine *(Passeriformes)* indet. | | 2 | | 2 |
| **birds/***Aves* | **8** | **59** | **27** | **97** |
| Σ | **19** | **87** | **34** | **110** |

*Samples – date of collection: 1 – Kojšovská hoľa, Volovské vrchy Mts, 26 June 2011 + 10 July 2011 (1040 m a.s.l.), 2 – Tupý vrch, Volovské vrchy Mts 18 June 2010 + 3 July 2010 + 12 June 2011 + 20 June 2011 + 26 June 2011 (1020 m a.s.l.), 3 – Stará Voda, Volovské vrchy Mts, 21 July 2013 + 11 June 2014 + 27 June 2014 (830 m a.s.l.).*

## Author details

Samuel Pačenovský[1*] and Alexander Kürthy[2]

1 State Nature Conservancy of Slovakia, Administration of Danube Lowlands Protected Landscape Area, Karloveská, Bratislava, Slovakia

2 State Nature Conservancy of Slovakia, Administration of Danube Lowlands Protected Landscape Area, Korzo Bélu Bartóka, Dunajská Streda, Slovakia

*Address all correspondence to: pacenovsky@vtaky.sk

## References


[1] Danko Š, Darolová A, Krištín A. Rozšírenie vtákov na Slovensku. Birds Distribution in Slovakia. Bratislava: VEDA; 2002. p. 686. [In Slovak with English summary]

[2] Kloubec B, Pačenovský S. Hlasová aktivita sýce rousného (*Aegolius funereus*) v jižních Čechách a na východním Slovensku: cirkadiánní a sezónní průběh, vlivy na její intenzitu. Buteo. 1996;**8**:5-22. [In Czech with English summary]

[3] Pačenovský S. Hniezdenie pôtika kapcavého (Aegolius funereus) na východnom Slovensku. Breeding of the Tengmalm's Owl (Aegolius funereus) in Eastern Slovakia. Zprávy MOS. 1992;**50**:27-32. [In Slovak with English summary]

[4] Pačenovský S, Shurulinkov P. Latest data on distribution of the pygmy owl (*Glaucidium passerinum*) in Bulgaria and Slovakia including population density comparison. Slovak Raptor Journal. 2008:91-106

[5] Pačenovský S, Šotnár K. Notes on the reproduction, breeding biology and ethology of the Eurasian pygmy owl (*Glaucidium passerinum*) in Slovakia. Slovak Raptor Journal. 2010;**4**(1):49-82. DOI: 10.2478/v10262-012-0046-y

[6] Pačenovský S. K medzidruhovým vzťahom Glaucidium passerinum, Strix uralensis *a* Strix aluco. On interspecific relationship among Glaucidium passerinum, Strix uralensis and Strix aluco. Tichodroma. 1995;**8**:61-73. [In Slovak with English summary]

[7] Šotnár K, Obuch J, Pačenovský S, Jarčuška B. Spatial distribution of four sympatric owl species in Carpathian mountain forests. Raptor Journal. 2020, 2020;**14**:1-13. DOI: 10.2478/srj-2020-0002

[8] Mošanský A. Faunistický prehľad vtákov (Aves) košickej aglomerácie (Východné Slovensko). Zborník Východoslovenského múzea v Košiciach, prírodné vedy. 1982;**22**:187-214. In Slovak with English summary]

[9] Danko Š. Doterajšie poznatky o kuvičkovi vrabčom (*Glaucidium passerinum*) vo východoslovenskom kraji. Haja. 1988;**3-4**(1986-1987):61-73. [In Slovak with English summary]

[10] Pačenovský S. Avifauna okresu Košice – mesto v rokoch 1979-81. Student work, biology olympiad. Secondary School Šmeralova. 1981;**9**:197. [In Slovak]

[11] Takáč K. Avifauna okolia Košíc. [In Slovak] Student Work, Secondary School Šrobárova 46, Košice; 1982. p. 52

[12] Pačenovský S Nové výskyty kuvička vrabčieho (Glaucidium passerinum) vo Volovských vrchoch a Čiernej hore (Slovenské rudohorie). Zborník Východoslovenského múzea, prírodné vedy; 1990. pp. 105-106. New occurrence of the Eurasian Pygmy Owl in Volovské and Čierna hora mountains. Proceedings of the East-Slovakian Museum: 105-106. [In Slovak with English summary]

[13] Pačenovský S. K aktivite a spotrebe potravy kuvička vrabčieho (Glaucidium passerinum). Tichodroma;**5**:17-21. Activity and food consumption of the Eurasian Pygmy Owl (Glaucidium passerinum). Tichodroma 5; 1993: 17-21. [In Slovak with English summary]

[14] Pačenovský S, Danko Š (eds.) et al. Birds of the Volovské mountains and their foothills. SOS/BirdLife Slovakia.

Bratislava; 2014. p. 180. [In Slovak with English summary]

[15] Pačenovský S, Kürthy A. Nové výsledky výskumu rozšírenia kuvička vrabčieho (*Glaucidium passerinum*) na východnom Slovensku a niekoľko poznámok k jeho etológii a hniezdnej bionómii. New results of research on distribution of the Eurasian Pygmy Owl in Eastern Slovakia and a few notes to its ethology and breeding bionomy. Buteo. 1991;**4**(1989):63-72. [In Slovak with English summary]

[16] Pačenovský S, Kürthy A. Zhodnotenie priebehu jesenného toku kuvička vrabčieho (Glaucidium passerinum) na základe hlasových prejavov. Zborník Východoslovenského múzea, prírodné vedy;**32-33**:55-68. Evaluation of the autumn-mating of Eurasian Pygmy Owl (Glaucidium passerinum) on the basis of its calls. Proceedings of the East-Slovakian Museum 32 – 33; 1992: 55-68. [In Slovak with English summary]

[17] Mikkola H. Owls of Europe. Calton: T & A D Poyser; 1983. p. 387

[18] Mikkola H, Sackl P. Pygmy Owl *Glaucidium passerinum*. In: The EBCC Atlas of European Breeding Birds. 1997. pp. 406-407

[19] del Hoyo J, Elliott A, Sargatal J. Handbook of the Birds of the World. Vol. 5. Barn-Owls to Hummingbirds. Lynx Edicions: Barcelona; 1999. p. 759

[20] Šotnár K, Pačenovský S, Obuch J. On the food of the Eurasian pygmy owl (*Glaucidium passerinum*) in Slovakia. Slovak Raptor Journal. 2015;**9**:115-126. DOI: 10.1515/srj-2015-0009

[21] Schmidt A, Pačenovský S. The breeding of pygmy owl (Glaucidium passerinum) on the Gömör–Tornai Karst. Aquila. 2011;**8**:87-96. [In Hungarian with English summary]

Chapter 5

# Owls (Strigiformes Wagler, 1830) in Bulgaria: Past and Present (A Review of the Fossil Record and Present Status of Recorded Species)

*Zlatozar Boev*

## Abstract

Two families, 8 genera, 12 species, and 13 subspecies of recent owls are recorded in Bulgaria. Two species (*Bubo scandiacus* and *Strix nebulosa*), established in the Pleistocene localities, disappeared from the country's recent avifauna. The southernmost limits of the European breeding ranges of three species (*Stix uralensis*, *Aegolius funereus*, and *Glaucidium passerinum*) pass through the territory of Bulgaria. Three species are endangered, 2—vulnerable, 1—threatened, and all the 12 species are protected by law. Earliest record of owls came from Gelasian (2.5 Mya) and Calabrian (1.6 Mya). Bone finds of two Early Pleistocene localities are incompletely identified (as *Asio* and *Athene*, respectively). The find of *Athene* sp. is determined as the oldest European record of that genus. Some anthropogenic factors that cause owl mortality are also presented.

**Keywords:** Pleistocene owls, Quaternary birds, paleoenvironment, avian bone remains, wildlife of Bulgaria, birds of Balkan Peninsula

## 1. Introduction

Owls (Strigiformes Wagler, 1830) in Bulgaria have always been held in special esteem. With their large round eyes, large heads, fascial disc, powerful and sharp claws, soft plumage and silent flight, they were the personification of strength, power, majesty, mystery and bad luck and death. Such beliefs are deeply rooted in the consciousness of broad circles of the population. Even today, many people believe that if they hear a Little Owl calling from the roof of the neighbor's house, there will soon be a funeral in that home.

In Bulgarian ethnography, there are no studies on the role of owls in the life of the population. Images or sculptures of owls in prehistoric and ancient art have so far not been found in the country, unlike some neighboring countries (Greece, for example). Past distribution of owl also remained out of special research, except that of [1–3].

At present, order Strigiformes includes between 213 [4] and 220 [5] recent species in two families—Strigidae Leach, 1820, and Tytonidae Ridgway, 1814. Palearctic



IntechOpen

fauna numbers 17 species [6], while in the Bulgarian avifauna 10 species are recorded. *Bubo scandiacus* (Linnaeus, 1758) and *Surnia ulula* (Linnaeus, 1758) are listed as likely to be found in the country [7], although none of them has been recorded so far. The influence of global warming in recent decades reduces the chances of establishment of these boreal species in Bulgaria.

Owls (large owls) are often one of the most active accumulators of animal remains (bones, teeth) in the cave deposits. Their role as taphonomic agents is highly appreciated by the paleozoologists, speleologist and ecologist [8]. The food of most species of owls in Bulgaria (based on pellets analysis) is relatively well studied, but the most numerous are the studies on the diet of the Eagle Owl. The largest number of food components were found in this species. It is believed that in Bulgaria, its number is among the highest in the Balkans and in Europe.

## 2. Material and methods

The chronostratigraphy follows [9] (Mya): Gelasian (GE) 2.588–1.800 (covering parts of the former Late Pliocene—Early Pleistocene); Calabrian (CA) 1.800–0.774 (Early Pleistocene); Chibanian (CH) 0.770–0.129 (Middle Pleistocene); Upper Pleistocene (UP) 0.129–0.0117 (Late Pleistocene); Greenlandian (GR) 0.0117–0.0082 (Early Holocene); Northgrippian (NO) 0.0082–0.0042 (Middle Holocene); and Meghalayan (ME) 0.0042–0.0001 (Late Holocene). Taxonomy is after [4, 5].

Abbreviations: Mya—million years ago, and Mnts—Mountains.

## 3. Past and present of Bulgarian owls

Species of both families, present in the Western Palearctic [6], have been recorded in the fossil (Pleistocene) and subfossil and recent (Holocene) avifauna of Bulgaria.

### 3.1 Tytonidae Ridgway, 1914

#### *3.1.1 Barn Owl (Tyto alba (Scopoli, 1769))*

Fossil/subfossil record: None.

Subspecies: *Tyto alba alba* (Scopoli, 1769) and *Tyto alba guttata* Brehm, 1931 [7].

Residence status: (*T. a. alba*): wintering [7], wintering, breeding [10]; (*T. a. guttata*): resident, vagrant, breeding, migratory [7].

Population number: (*T. a. alba*): no data; single pairs (*T. a. guttata*): 1500–1700 pairs [11], 500–1000 pairs [12], 600 pairs [10].

Conservation status: (*T. a. alba*): protected [12], vulnerable [10]; (*T. a. guttata*): protected [12], vulnerable [10].

Peculiarities: Both European subspecies are spread in the country.

### 3.2 Strigidae Wagler, 1830

#### *3.2.1 Eurasian Pygmy Owl Glaucidium passerinum (Linnaeus, 1758)*

Fossil/subfossil record: UP: [13–15]. The find from the Devetashka Cave is the fifth World fossil record of this species [13].

Subspecies: *Glaucidium passerinum passerinum* (Linnaeus, 1758) [7].
Residence status: resident, breeding [7, 12].
Population number: 100–120 pairs [11], 30–120 pairs [12], 150–200 pairs [16].
Conservation status: threatened [12], endangered [10].
Peculiarities: The southern border of the breeding range of the species passes through the territory of the country. Bulgarian populations are refugial montane and inhabit old-growth prime coniferous forests [17, 18]. The largest compact prime habitat on the Balkan Peninsula is preserved in the Rhodopes Mnts. [19].

#### *3.2.2 Little Owl (Athene noctua (Scopoli, 1769))*

Fossil/subfossil record: UP, GR, NO, ME: [1, 2, 14, 15, 20–23].
Subspecies: *Athene noctua indigena* Brehm, 1855 [7].
Residence status: resident, wintering, vagrant, breeding [7, 12].
Population number: 16,000–18,000 pairs [11], 7500–10,000 pairs [12].
Conservation status: protected [12].
Peculiarities: 90% of the Bulgarian Little owls breed and winter in human settlements [24].

#### *3.2.3 Athene F. Boie, 1822 sp.*

Fossil/subfossil record: GE: [25–27]. This is the oldest record of a strigiform bird in Bulgaria. It is dated ca. 2.5 Mya. The only find (incomplete sternum) came from the Early Pleistocene (Middle Villafranchian, MN 17) locality near Dolno Ozirovo village (Montana Region; NW Bulgaria), known as Varshets locality. The find is evaluated as the oldest so far European record of that genus in Europe [28].

#### *3.2.4 Boreal (Tengmalm's) Owl (Aegolius funereus (Linnaeus, 1758))*

Fossil/subfossil record: UP, ME: [14, 15].
Subspecies: *Aegolius funereus funereus* (Linnaeus, 1758) [7].
Residence status: resident, vagrant, breeding [7, 12].
Population number: 1100–1200 pairs [11], 600–900 pairs [12].
Conservation status: vulnerable [10, 12].
Peculiarities: The Pleistocene locality of the species in the Devetashka Cave falls outside the modern breeding range. The southern border of the breeding range of the species passes through the territory of Bulgaria and Greece. Bulgarian populations are refugial montane and inhabit old-growth prime coniferous forests [17, 18]. The largest compact prime habitat on the Balkan Peninsula is preserved in the Rhodopes Mnts. [19].

#### *3.2.5 Eurasian scops owl (Otus scops (Linnaeus, 1758))*

Fossil/subfossil record: UP, GR, NO: [13, 23, 29, 30].
Subspecies: *Otus scops scops* (Linnaeus, 1758) [7].
Residence status: breeding, migratory [7].
Population number: 12,000–14,000 pairs [11], 6000–9000 pairs [12].
Conservation status: protected [12].
Peculiarities: Some individuals winter in the country [31]. A singing male recorded on 12.02.2014 in a park in Sofia City [32].

### 3.2.6 Long-eared Owl (*Asio otus* (Linnaeus, 1758))

Fossil/subfossil record: UP, GR, NO, ME: [13, 23, 29, 33].
Subspecies: *Asio otus otus* (Linnaeus, 1758) [7].
Residence status: wintering, breeding [7, 12].
Population number: 12,000–14,000 pairs [11], 3000–5000 pairs [12].
Conservation status: protected [12].
Peculiarities: In winter, migratory specimens from the north form numerous aggregations in forest areas in the country, where they usually stay for several weeks during the coldest periods.

### 3.2.7 Short-eared Owl (*Asio flammeus* (Pontoppidan, 1763))

Fossil/subfossil record: UP, NO: [13, 23].
Subspecies: *Asio flammeus flammeus* (Pontoppidan, 1763) [7].
Residence status: wintering, resident, breeding [7, 12].
Population number: 3–6 pairs [11], 0–6 pairs [12].
Conservation status: threatened [12].
Peculiarities: The southern border of the breeding range of the species passes through the territory of the country. Last breeding recorded in July 2009 [34].

### 3.2.8 *A. otus* (Linnaeus, 1758)/*A. flammeus* (Pontoppidan, 1763)

Fossil/subfossil record: UP: [14]. The finds (20 items) originate from the Late Wurm (Middle Paleolithic, ca. 70,000 BP) deposits of the Devetashkata Cave near Devetaki village (Lovech Region, CN Bulgaria). It is the largest Bulgarian Cave and the richest avian paleontological locality in Bulgaria. Most of the finds represent pedal phalages, ulnare bones, or bone fragments of immature individuals that could not be reliably identified up to species level. Therefore, it is preferable to leave the determination open, i.e., until genus level (*Asio*) with an assumption of one of two mentioned species.

### 3.2.9 Tawny Owl (*Strix aluco* Linnaeus, 1758)

Fossil/subfossil record: CA, UP, NO, ME: [1, 2, 13–15, 20, 23, 35–37].
Subspecies: *Strix aluco aluco* Linnaeus, 1758 [7].
Residence status: resident, breeding [7].
Population number: 10,000–12,000 pairs [11] 4000–8000 pairs [12].
Conservation status: protected [12].
Peculiarities: Individuals from both phases, gray (**Figure 1**) and brown (**Figure 2**), have been found in Bulgaria.

### 3.2.10 Ural Owl (*Strix uralensis* Pallas, 1771)

Fossil/subfossil record: None.
Subspecies: *Strix uralensis liturata* Tengmalm, 1793 [7].
Residence status: resident, vagrant, breeding [7, 12].
Population number: 150–200 pairs [11], 20–60 pairs [12].
Conservation status: threatened [12], endangered [10].

**Figure 1.**
*Tawny owl (Strix aluco aluco), gray phase. Orsoya village (Montana Region, NW Bulgaria), 26.04.2020. Photograph: Nikolay Karaivanov.*

Peculiarities: The southern border of the breeding range of the species passes through the territory of the country. The Bulgarian population is a relict of the last glaciations [38].

*3.2.11 Great gray Owl (Strix nebulosa Forster, 1772)*

Fossil/subfossil record: UP: [13, 14, 39].
Residence status: not recorded in the Holocene [7].
Population number: none [7, 12].
Conservation status: none.
Peculiarities: It is possible that the species will be established in Bulgaria during the eruptive movements in the winter period.

*3.2.12 Eurasian Eagle-Owl (Bubo bubo (Linnaeus, 1758))*

Fossil/subfossil record: CA, UP, GR, NO, ME: [1, 2, 14, 15, 23, 40–43].
Subspecies: *Bubo bubo bubo* (Linnaeus, 1758)) [7].
Residence status: resident, breeding [7, 12].
Population number: 120–150 pairs [44], 650–700 pairs [11], 420–490 pairs [12].
Conservation status: threatened [12], endangered [10].
Peculiarities: The species' population obviously increases at least triple in the last 3–4 decades, despite the decades-long incidents of shooting specimens by unenlightened hunter-poachers (**Figures 3** and **4**).

*3.2.13 Snowy Owl (B. scandiacus (Linnaeus, 1758))*

Fossil/subfossil record: UP: [13, 14, 45].
Residence status: not recorded in the Holocene; probable [7].
Population number: none [7, 12].
Conservation status: none.

**Figure 2.**
*Tawny owl (Strix aluco aluco), brown phase. Ofeliite Locality, Vitosha Mnt. (CW Bulgaria), 22.06.2020. Photograph: Lyubomir Hristov.*

**Figure 3.**
*A shot Eagle owl (Bubo bubo bubo). Near Reselets village (Pleven Region, NW Bulgaria). 16.07.1977. Photograph: Zlatozar Boev.*

Peculiarities: It is possible that the species will be established in Bulgaria during the winter period.

## 4. Discussion

Based on the occurrences as rare winter vagrants in the neighboring Balkan countries, [7] (1990) list two additional species as probable for the Bulgarian avifauna—*S. ulula* (subspecies *S. u. ulula* Linnaeus, 1758.) and *B. scandiacus*. In the last more than three decades, there have been no sightings of both species, but we do not rule out the possibility of their records. Both species have a propensity for non-periodic long winter migrations to the south.

**Figure 4.**
*A shot Eagle owl (Bubo bubo bubo). Near Town of Belogradchik (Vidin Region, NW Bulgaria). 01.10.1993. Photograph: Zlatozar Boev.*

*B. scandiacus* is already established in the country, albeit in the Late Pleistocene, and for *S. ulula*, its Late Pleistocene presence has recently been established in the Eastern Palearctic as far south as even North Vietnam [46].

Although all owls in Bulgaria are protected by law, numerous are the hazards that still cause their death. The road kills from traffic have the most significant impact. In the open plain and lowland landscapes of Southern Bulgaria, Barn owls especially often die like this. However, they are not the only victim of the roads among the owls. Some species preferring woodlands and forest habitats also die in the same way on mountain roads (**Figure 5**).

At night, owls easily find the still warm bodies of killed or injured birds and small mammals on the roadway and swoop down on easy prey. In such situations on the road, blinded by the powerful headlights of cars, they are run over. Barn owls often die like this every year in southern Bulgaria.

**Figure 5.**
*A Tawny owl (Strix aluco aluco) run over on the road. Studenets Natura 2000 Special Protected Area (NC Bulgaria. Photograph: 12.11.2009. Zlatozar Boev.*

**Figure 6.**
*A Little owl (Athene noctua indigena) died from the 230-volt electric voltage on open power lines in the village of Archar (Vidin Region, NW Bulgaria). Photograph: 12.11.2009. Zlatozar Boev.*

A special survey of a 68-kilometer section of the Thrace highway between the cities of Pazardzhik and Plovdiv found that seven individuals of *A. otus* and five individuals of *T. alba* died in 20 days [47].

Some other relatively rare factors also contribute to owl mortality. In the urban areas (even rural), owls are sometimes injured and being killed by the electricity power as a result of flying into overhead wires (**Figure 6**). Unfortunately, not only small owls such as Little owls, but also the largest Eagle owls die from an electric shock from a short circuit. Such a case is presented by [48], making a proposal to replace electric poles or secure them with respect to birds. More than 30 years have passed since then, but today the situation is not much different.

## 5. Conclusions

Two owl species (*B. bubo* and *S. aluco*) are established in the Calabrian, and a genus (*Athene*) even in the Gelasian. Data from Bulgaria confirm that *B. bubo* and *S. aluco* coexisted with the Paleolithic man at least since 1.6 Mya. Although hominine remains have not been found in the Gelasian bird localities in Bulgaria, it is beyond any doubt that the owls and the first people in Bulgaria and the Balkans shared their cave dwellings. The present study is the first attempt to summarize available information on the composition, chronostratigraphic distribution, current conservation, and residental status and threats of owls in Bulgaria. Although scarce, these are also the first data on the past of owls in the country.

The Quaternary Bulgarian strigiform avifauna is rich and diversified. Two families, 8 genera, 12 species, and 13 subspecies of recent owls are recorded in the country's nature in the last over 2 Mya. Bone finds of two Early Pleistocene localities are incompletely identified (as *Asio* and *Athene*, respectively). Two species (*B. scandiacus* and *S. nebulosa*) disappeared from the country's recent avifauna. The southernmost limits of the breeding ranges of three species (*Stix uralensis*, *A. funereus*, and *G. passerinum*) pass through the territory of Bulgaria, which lies on the southern periphery of their ranges. Three species are endangered, 2—vulnerable, 1—threatened, and all the 12 species are protected by law. Only one species (*T. alba*) is represented by two subspecies in the Bulgarian fauna. Earliest record of owls came from Gelasian (2.5 Mya) and Calabrian (1.6 Mya). The find of *Athene* sp. is determined as the oldest European record of that genus.

Although not abundant, the fossil/subfossil record of owls is highly intriguing and promising field of paleozoological research.

## Acknowledgements

The author thanks Dr. Nikolay Karaivanov and Mr. Lyubomir Hristov for the provided photographs of *S. aluco*.

## Conflict of interest

The author declares no conflict of interest.

## Author details

Zlatozar Boev
National Museum of Natural History, Bulgarian Academy of Sciences, Sofia, Bulgaria

*Address all correspondence to: boev@nmnhs.com; zlatozarboev@gmail.com

## References

[1] Boev Z. Raptors and owls (Aves: Falconiformes et Strigiformes) in the Archaeological Record of Bulgaria. Historia Naturalis Bulgarica. 1996;**6**:83-92

[2] Boev Z. The holocene avifauna of Bulgaria (A review of the ornitho-archaeological studies). Historia Naturalis Bulgarica. 1996;**6**:59-81

[3] Boev Z. Birds in everyday life and art in Bulgaria (Thracian and Roman periods). Historia Naturalis Bulgarica. 2018;**27**:3-39

[4] Dickinson EC, Remsen JV. The Howard & Moore Complete Checklist of the Birds of the World. 4th. ed. Vol. 1. Eastboume, U.K.: Aves Press; 2013. p. 461

[5] del Hoyo J, Collar NJ. HBW and BirdLife International Illustrated Checklist of the Birds of the World. Volume 1: Non-passerines. Barcelona: Lynx Edicions; 2014. p. 903

[6] Cramp S. Handbook of the Birds of Europe the Middle East and North Africa. The Birds of the Western Palearctic, Vol. IV. Terns to Woodpeckers. Vol. 4. Oxford Univiversity Press; 1989. p. 960

[7] Simeonov S, Michev T, Nankinov D. Fauna of Bulgaria. Vol. 20. Sofia: Aves, Part I., Publishing House of the Bulgarian Academy of Sciences; 1990. p. 351

[8] Andrews P. Owls, Caves and Fossils. Predation, preservation and accumulation of small mammal bones in caves, with an analysis of the Pleistocene cave faunas from Westbury-sub-Mendip, Somerset, UK. London: Published by The Natural History Museum; 1990. p. 231

[9] Cohen KM, Finney SC, Gibbard PL, Fan J-X. The ICS international chronostratigraphic chart. Episodes. 2013;**36**:199-204

[10] Golemanski V. (Editor-in-Chief) Red Data Book of the Republic of Bulgaria. Volume 2. Animals. Sofia: IBEI—BAS & MOEW; 2011. p. 372

[11] Nankinov D et al. Breeding Totals of the Ornithofauna in Bulgaria. Green Balkans: Plovdiv; 2004. p. 32

[12] Iankov P, editor. Atlas of Breeding Birds in Bulgaria. Bulgarian Society for the Protection of Birds, Conservation Series, Book 10. Sofia: BSPB; 2007. p. 679

[13] Boev Z. Neogene and Quaternary birds (Aves) from Bulgaria. Bulgarian Acadademy of Sciences, National Museum of Natural History D. Sci. Thesis, Sofia, Volume I. Basic Part. 243 pp.; Volume II. Supplement 1—Figures, 135 pp.; Volume II. Supplement 2—Tables. 1999. p. 108 (In Bulgarian)

[14] Boev Z. Birds over the mammoth's head in Bulgaria. In: Cavaretta G, Gioia P, Mussi M, Palombo MR, editors. The World of Elephants. Roma: Proceedings of the 1st International Congress; 2001. pp. 180-186

[15] Boev Z. Pleistocene avifaunas of Bulgaria: A brief review. Historia Naturalis Bulgarica. 2006;**17**:95-107

[16] Pačenovský S, Shurulinkov P. Actual knowledge on distribution of the Pygmy Owl (*Glaucidium passerinum*) in Bulgaria and Slovakia with population density comparison. Slovak Raptor Journal. 2008;**2**:89-106

[17] Shurulinkov P, Stoyanov G. Some new findings of Pigmy Owl *Glaucidium passerinum* and Tengmalm's Owl *Aegolius funereus* in western and southern Bulgaria. Acrocephalus. 2006;**27**(128-129):65-68

[18] Nikolov B, Zlatanov T, Groenb T, Stoyanovc S, Hristova-Nikolova I, Lexere M. Habitat requirements of Boreal Owl (*Aegolius funereus*) and Pygmy Owl (*Glaucidium passerinum*) in rear edge montane populations on the Balkan Peninsula. Avian Research. 2022;**13**(100020):1-8

[19] Shurulinkov P, Ralev A, Daskalova G, Chakarov N. Distribution, numbers and habitat of Pigmy Owl *Glaucidium passerinum* in Rhodopes Mts (S Bulgaria). Acrocephalus. 2007;**28**(135):161-165

[20] Boev Z, Beech M. The bird bones. In: Poulter AG, editor. Nicopolis ad Istrum. A Late Roman and Early Byzantine City. The Finds and the Biological Remains. London: Oxbow Books. The Society of Antiquaries of London; 2007. p. 242-253+307-318

[21] Boev Z. Middle and late holocene birds from the eastern Upper Thracian Plane (S Bulgaria). Historia Naturalis Bulgarica. 2004;**16**:123-132. (In Bulgarian, English summary)

[22] Manhart H. Die vorgeschichtliche Tierwelt von Koprivec und Durankulak und anderen prahistorischen Fundplatzen in Bulgarien aufgrund von Knochcnfunden aus archaologischen Ausgrabungen. Documenta naturae. 1998;**116**:1-353

[23] Mitev IV. Subfossil finds of birds and mammals in accumulations of the food of Eagle Owl (*Bubo bubo* (L., 1758)) (Aves: Strigiformes) from the valley of Rusenski Lom river. Historia Naturalis Bulgarica. 2006;**17**:137-151

[24] Ignatov A, Popgeorgiev G 2021. Recent and historical distribution of Little Owl (*Athene noctua*) in Bulgaria.—Airo, 2021;**11**:216-222

[25] Boev Z. Neogene avifauna of Bulgaria. In: Zhou Z, Zhang F, editors. Proceedings of the 5th Symposium of the Society of Avian Palaeontology and Evolution, Beijing. Beijing: Science Press; 2002. pp. 29-40

[26] Boev Z. Neogene avifaunas of Bulgaria (a brief review). In: Bakardjieva N, St. Chankova B, KrastanovS, editors. Gateva (Compilers). Evolution and Ecology—2007. Union of the Scientists of Bulgaria. Sofia: 3rd National Seminar. Proceedings; 2007. pp. 26-35

[27] Boev Z. Paleobiodiversity of the Vrachanska Mountains in the Villafranchian: a case study of the Varshets (Dolno Ozirovo) Early Pleistocene locality of fossil fauna and flora. In: Bechev D, Georgiev D, editors. Faunistic Diversity of Vrachanski Balkan Nature Park, ZooNotes, Supplement 3. Plovdiv: Plovdiv University Press; 2016, 2016. pp. 299-323

[28] Mlĭkovský J. Cenozoic Birds ofthe World. Part 1, Europe. Praha: Ninox Press; 2002. p. 406

[29] Mitev I, Boev Z. Food spectrum of the Eagle Owl (*Bubo bubo* (L., 1758)) (Aves: Strigiformes) from two Holocene localities in NE Bulgaria. Historia Naturalis Bulgarica. 2006;**17**:153-165. (In Bulgarian, English summary)

[30] Mitev I. Comparative analysis of the food spectrum of the Eagle owl (*Bubo bubo* (Linnaeus, 1758)) in two localities from the Northeast Bulgaria. In: Boev Z, editor. Ivan Mitev. Selected Works. Vol. 1. Bulgarian Nature. Sofia: Logis Publishing House; 2016. pp. 118-154

[31] Nankinov D. Ptitsata Chuh-chuh.—Lov i ribolov. 1991;**1**:8-9 (In Bulgarian)

[32] Boev Z. Data on the study of bird distribution in Bulgaria in the last 45 years (1976-2020). Bulletin of the Natural History Museum, Plovdiv. 2021;**6**:1-45

[33] Boev Z. Late Pleistocene Avifauna of the Razhishkata Cave, Western Bulgaria. Historia Naturalis Bulgarica. 2000;**12**:71-87

[34] Dimchev I, Mladenov V. Gnezdene na blatnasova (*Asio flammeus*) v Atanasovsko ezero. Za ptitsite. 2010;**1**:15. (in Bulgarian)

[35] Boev Z. The birds of the Roman town of Nicopolis ad Istrum (2nd–6th c. AD) at Nikjup, Lovech Region. Historia Naturalis Bulgarica. 1991;**3**:92-102. (In Bulgarian, English summary)

[36] Boev Z. Avian remains from the Late Neolithic settlement of Hadzhidimitrovo (Yambol Region, SE Bulgaria). In: Petrova V, editor. The Late Neolithic settlement of Hadzhidimitrovo (Yambol Region, SE Bulgaria). Sofia: Prof. Marin Drinov Academic Publishing House; in press

[37] Boev Z, Iliev N. Les oiseaux et leur importance pour les habitants de Veliki Preslav (IXe-X-e s.). Arheologiya, BAS. 1991;**3**:43-48. (In Bulgarian, French summary)

[38] Spiridonov G, Spassov N, Mileva L. New data on the distribution of the Ural owl (*Strix uralensis*) and the Boreal owl (*Aegolius funereus*) in Bulgaria. In: National Theoretical Conference of the Conservation and Reproduction of the Environment. Vol. 1. Slanchev Bryag; 1982. pp. 341-343. (in Bulgarian)

[39] Boev Z, Mikkola H. First Pleistocene record of Great Grey Owl (*Strix nebulosa* Forster, 1772) in Bulgaria. Comptes rendus de l'Académie bulgare des Sciences. 2022;**75**(5):680-685

[40] Bochenski Z. Aves. In Excavations in the Bacho Kiro Cave (Bulgaria). Final report. In: Kozlowski J, editor. Warszawa: Panstwowe Wydawnictwo Naukowe; 1982. pp. 31-38

[41] Boev Z. Neolithic birds from the Prehistoric settlement at Kazanluk. Historia Naturalis Bulgarica. 1993;**4**:57-67. (in Bulgarian, English summary)

[42] Boev Z. The Upper Pleistocene Birds. In: Kozlowski JK, Laville H, Ginter B, editors. Temnata Cave. Excavations in Karlukovo Karst Area, Bulgaria. Vol. 1.2. Cracow: Jagellonian Univ. Press; 1994. pp. 55-86

[43] Boev Z. Birds of the Roman settlement Arbanas—1 near town of Pernik. Historia Naturalis Bulgarica. 1997;7:28. (in Bulgarian)

[44] Simeonov S, Michev T. On the present distribution and abundance of the eagle owl (*Bubo bubo* (L.) in Bulgaria). Ekologiya. 1985;**15**:60-65. (In Bulgarian, English summary)

[45] Boev Z. First fossil record of the Snowy Owl *Nyctea scandiaca* (Linnaeus, 1758) (Aves: Strigidae) from Bulgaria. Historia Naturalis Bulgarica. 1998;**9**:79-86

[46] Boev Z. Late Pleistocene and Early Holocene Birds of Northern Vietnam (Caves Dieu and Maxa I, Thanh Hoa Province)—Paleornithological Results of the Joint Bulgarian-Vietnamese Archaeological Expeditions, 1985-1991 (Paleoavifaunal Research). Quaternary. 2022;**5**:31. DOI: 10.3390/quat5030031

[47] Kambourova-Ivanova N, Koshev Y, Popgeorgiev G, Ragyov D, Pavlova M,

Mollov I. Nedialkov 2012. Effect of traffic on mortality of amphibians, reptiles, birds and mammals on two types of roads between pazardzhik and plovdiv region (Bulgaria)—Preliminary results. Acta Zoologica Bulgarica. 2012;**64**(1):57-67

[48] Ts P. Suicide or murder of the Eagle Owls. Zashtita na Pritodata. 1990; **10-11**:70-71. (in Bulgarian)

Chapter 6

# Queens of the Night, the Owls of Iraq and Syria - Species, Current Distribution, and Conservation Status

*Omar F. Al-Sheikhly and Ahmad E. Aidek*

## Abstract

The owls in the Arabian culture are a symbol of a bad omen and the embodiment of evil spirits that are roaming in deserts and abandoned places at night to impersonate the wandering nomads. In other stories, for centuries, the Bedouins in the deserts of Iraq and Syria are considered owls as the flying spirits of dead warriors who seek revenge when they hoot or as a sign of destruction. For decades, owls with their fierce and protruding eyes resembled satanic birds; therefore, they are persecuted to be used in sorcery whenever and wherever possible. The old beliefs and traditional knowledge of the Iraqi-Syrian local communities are still an "unsolved" critical issue that interferes with owls' conservation efforts in the Middle East. There are 10 owl species occurring in Iraq and Syria; yet, their updated status and zoogeographical extent are still not fully explored. In this chapter, an annotated checklist of owls in Iraq and Syria, their historical and current distribution, and conservation status are provided.

**Keywords:** birds of prey, human-wildlife conflict, species distribution, Strigiformes of the Middle East, traditional knowledge

## 1. Introduction

### 1.1 Iraq

Iraq is situated between latitudes 29° to 38° N and longitudes 39° to 49°, a small area lies west of 39°, and spans over 437,072 km$^2$. It borders Turkey to the north, Iran to the east, Jordan to the west, Saudi Arabia to the south and southwest, and a narrow coastline with Kuwait to the south. Iraq is sharing a vast transboundary monotonic habitat, including desert, arid steppes, and plateau (Al-Badiyah), with Syria in the west and northwest.

In Iraq, the first attempt to study the owl fauna of the country was conducted by British ornithologists who visited Iraq as members of the British military campaign in the 1920s. They provided detailed information on taxonomic groups, morphological remarks on species/subspecies occurrence, and migration patterns. However, the status of many owl species in Iraq was not fully discovered due to a lack of subsequent *in situ* ornithological surveys with a significant lack of observations between the 1980s–1990s due to conflicts in the region. In the 1920s, avifaunal observations were compiled in 1922 by Ticehurst et al. [1] who reported seven owl species from Mesopotamia (territory of Iraq and Southwestern Iran) with additional observations added in 1926 by Ticehurst et al. [2]. In the 1950s, an ornithological survey was conducted by Moore and Boswell [3], who reported a total of five owl species in the country. Later on, scattered owl observations were made by several European ornithologists in Iraq, for example, Sage [4, 5], Marchant [6, 7], and Scott and Carp [8] during the subsequent years (see list of species—Iraq). Nevertheless, in the major avifauna monograph of Iraq, which was made in 1960–1962 by Allouse [9], a total of nine owl species were reported from Iraq.

In regard to the Scops owls in Iraq, Allouse [9] mentioned that *Otus* spp., including records of both Eurasian Scops Owl *Otus scops* and Pallid Scops Owl *O. brucei*, further indicating those observations from Iraq were grouped under *Otus* group taxa but with doubts. An attempt to study the avifauna of the steppes and arid regions of Central Iraq was made 38 years later by Al-Dabbagh [10], who reported the occurrence of three owl species, the Western Barn Owl *Tyto alba*, Eurasian Scops Owl, and Eurasian Eagle Owl *Bubo bubo*.

In recent ornithological studies, that is, Porter et al. [11] listed nine owl species from Iraq excluding the Pharaoh Eagle Owl *Bubo ascalaphus*; although, the occurrence of this arid land-dowelling species was confirmed in 1960 in the desert of Western Iraq by Vaurie [12] and the updated geographical distribution and breeding status in Iraq was confirmed later on by Al-Sheikhly [13] and Al-Sheikhly et al. [14]. Furthermore, the last paper [14] comprehensively discussed the geographical distribution and taxonomic status of the Eurasian Eagle Owl *B. bubo* ssp. complex in Iraq in comparison with morphological traits of the recently discovered Pharaoh Eagle Owl in Southwestern Iran, indicating that more research is required to determine the Eurasian Eagle Owl subspecies existing in Iraq and Iran.

In regard to the *Asio* spp., both Northern Long-eared Owl (thereafter only Long-eared Owl) *Asio otus* and Short-eared Owl *Asio flammeus* are reported in Iraq by Ticehurst et al. [1]. They questioned the validity of the occurrence of the pale eastern race *A. f. leucopsis*; as a specimen obtained from Qalet Saleh, which was a pale bird, while the rest were all as dark as West European races. Seasonal morphological variations in Short-eared Owl may also occur in Iraq. Ticehurst et al. [1] mentioned that there are seven specimens in the British Museum obtained from Fao in Southern Iraq, three were light birds (October and March) and four are dark birds (one October and the rest bear no date). Therefore, further research is needed to reveal the taxonomic status of this genus in Iraq.

Salim et al. [15] listed 11 owl species from Iraq considering the Lilith Owlet *Athene (noctua) lilith* as a distinct species. Regarding the Little vs. Lilith owls' contention in literature, Ticehurst et al. [1] mentioned that Little owls from Iraq are somewhat varied in the color of the upperparts even within the same locality; they are too dark, not pale sandy enough above for those *A. n. lilith* (from Palestine), and certainly are not like the North African Little Owl *A. n. glaux* (from Egypt). Iraqi Little owls pretty much resembled Hutton's Little Owl *A. n. baetriana* from Kandahar (Afghanistan). Moreover, they gave morphological remarks between *A. n. lilith* vs. *A. n. baetriana*, as the feathering on the toes varies; on the whole, *Lilith* is less feathered than *bactriana* and more so than *Lilith* but some are as fully feathered as the former usually is, while skins from Kandahar in winter, the toes have little more than bristles on them. Apart from individual variation in color, the season also makes a difference, *bactriana* in spring is noticeably paler than in autumn. On the underside, these Mesopotamian birds have the strikes not so yellowish-red as in *lilith*. The North African Little Owl *A. n. glaux* had been recorded from the Iranian Karun district and both *glaux* and *bactriana* are residents in the Zagros Mountains [1]. Reports also revealed that *glaux* commonly found in Babylon (Central Iraq) while records from Mosul (Northern Iraq) indicated that adult *bactriana* birds look quite like *glaux*. A mummified specimen obtained in winter at Urfa in Southern Turkey was regarded as *glaux,* while the type of *lilith* reported from the upper Euphrates at Der-ez-Zor (between Deir and Aleppo), which is also recorded from Southwestern Persia. Speculations on the taxonomy of these ssp. in Iraq were raised as little owls from Shustar in the south of Samarra in Northern Iraq were of the race *bactriana,* while no specimens were certainly identified as *glaux* obtained from Iraq (see [1]). Despite its enigmatic status in Iraq; however, this taxon is considered as a subspecies of Little Owl in the taxonomic revision by OSME [16].

### 1.2 Syria

Syria (the Syrian Arab Republic) is located on the eastern edge of the Mediterranean Sea between latitudes 32° 19′ and 37° 30’ N and longitudes 35° 45′ and 42° 25′ E and spans over 185,180 $km^2$. The country borders Turkey to the north, Iraq to the east, Jordan to the south, and shares a short coastline with Lebanon and borders the Mediterranean Sea to the west. This wide range of transboundary terrestrial habitats awards both Syria and Iraq with a significant diversity of fauna and flora, the owl fauna being not an exception.

In Syria, the owl fauna has not been fully discovered; however, several kinds of research related to systematics, distribution, or species diversity of small mammal prey taken by owls, or to determine their role as biological control agents on rodent pests were conducted during the last three decades (Shehab and Johnson [17]; also see Literature records—Syria).

Among the owl species in Syria, the Western Barn Owl was comprehensively studied as it seems to be the ideal species to investigate the remains of mammalian prey in owl pellets [18, 19]. It is the most widespread common owl species in Syria and its distribution and ecology were intensively discussed in [17, 20].

Due to their secretive behavior and enigmatic status in Syria, other owl species, such as Pallid Scops Owl, Eurasian Scops Owl, Eurasian Eagle Owl, and Pharaoh Eagle Owl were of great interest to European and local ornithologists. The Pallid Scops Owl is considered among rarities in Syria, with only three records reported [17]. Breeding evidence was reported in 1924 from Aleppo in Northern Syria by Clarke [21], a single sighting was reported from Tall Shekh Hamad at the Khabur by Baumgart et al. [22], and from Sabkhat al-Jabbul in Halab and Euphrates valley by Evans [23]. The status of the Eurasian Scops Owl is enigmatic, unverified records were made form Ras Al-Baseet, Qatana [22]. The only documented record came from a frozen hunted specimen at the local animal market in Damascus in 2005 [17].

The status of the Eurasian Eagle Owl in Syria is not been fully explored; however, the species occurrence in the country was reported by Obuch [24], Shehab [25], Shehab and Mamkhair [26], Murdoch et al. [27], Benda et al. [28], Serra et al. [29], Shehab and Johnson [17], and Bowler [30]. The Pharaoh Eagle Owl was confirmed breeding in the wadies of Eastern Syria [29, 31].

Despite its broad range, additional few studies related to the Little Owl *A. noctua* as a specialist to take arthropods' prey were conducted by Obuch and Kristín [32] and Shehab et al. [33].

Shehab and Johnson [17] mentioned that there are seven owl species in Syria, indicating that further survey may determine whether an eighth species, such as Brown Fish Owl *Ketupa zeylonensis,* should be excluded or not. This owl is now considered an extremely rare species in the country [34]. Anyhow, recent photos have been published on local social media, indicating that this species may still be present in extreme Northern Syria (Ahmad Aidek to Omar Al-Sheikhly pers. comm. 2022). This is interesting, especially when there is a resident Brown Fish Owl population in Southern Turkey adjacent to Syrian borders [35].

Nevertheless, Baumgart et al. [22] reported a total of 10 owl species in Syria, including the first confirmed occurrence of the Eurasian Scops Owl in the country. However, they did not provide further details supporting the presence of the Short-eared Owl, Hume's Owl *Strix butleri,* and Long-eared Owl [22]. The occurrence of two owl species, the Long-eared Owl and Tawny Owl *Strix aluco*, in Syria was reported by Obuch [24] and Benda et al. [28]. The breeding of the Long-eared Owl was confirmed in Northwestern Syria by Manners and Diekmann [36]. However, Shehab and Johnson [17] indicated that the Long-eared Owl is very rare in Syria, with only four records reported. They also reviewed the juvenile Long-eared Owl record from Dana village near Bab Al-Hawa, 40 km west of Halap made by Manners and Diekmann [36] and reidentified that juvenile owl as a Eurasian Eagle Owl later on. In addition, Murdoch and Betton [34] had reported 10 owl species in Syria, indicating that Hume's Owl has not been reported yet in the country but could well occur near the Jordanian borders; therefore, further investigation is required.

There are 10 confirmed owl species occurring in Iraq and Syria; yet, their updated status and zoogeographical extent are still not completely known. In this chapter, an annotated checklist of owls in Iraq and Syria, their historical and current distribution, and conservation status along with their challenging conservation issues, are provided. Owl species are listed according to the OSME Region List of Bird Taxa [16]. The species' conservation status is listed according to the *International Union for Conservation of Nature* (IUCN) Red List. The status of each species is based on Salim et al. [15] for Iraq and Murdoch and Betton [34] for Syria. In addition, the literature species status, breeding status, and distribution maps of owl species were updated based on recent records/observations made by the authors combined with other observations made by several field ornithologists (see acknowledgments) throughout Iraq and Syria during 2007–2022. Literature and recent owl records were placed on an updated distribution topographical map for Iraq and Syria, where black dotes represent previous literature records and white dotes the recent records.

## 2. Owls in Arabian tradition of Iraq and Syria

The symbolics of animal species have some contradictory representations in the literature of different ethnic groups and cultures. These symbols could represent peace, desire, and friendship, while others could be a sign of hostility in some people's beliefs [37]. One of the animals with the greatest contradictions is the owl. An animal that, despite its many denunciations in some cultures, has its own position, albeit a negative one. Owls have been known by the Arabian culture as those wild birds with two horns on top of the large head, broad face, great fierce piercing eyes, curved beaks, and powerful legs equipped with sharp talons. Owls mostly inhabit ruins and caves in deserts and wadies, taking their prey at night with secretive behavior, noticed when heard with their heart-taking hoots rather than being visually seen.

In Islamic belief, the owl came from the birds of prey characterized by their hooked beaks, rounded heads, and very short necks. The eyes are immobile and large surrounded by feathers attached to them. They fly slowly and have an accurate hearing sense [37, 38].

The Arabian zoologist, Ibn Musa Al-Damiri (1370–1405) mentioned in his Life of Animals (*Ḥayāt al-ḥayawān al-kubrā*) book that owls at night have more power, dominant and hostile on other creatures; they are nocturnal, attacking other birds' nests' prey on their eggs and nestlings. They are also known for their secretive loneliness and their hostility toward crows as intrinsic behavior. Based on different Arabian tribal distribution, owls are known by several Arabic names, for example, *Boom*, *Umm al-Sharab*, *Umm al-Sabian*, *Umm al-Saeed, Abu Al-Manhal*, *Abu Malik,* and others. Allouse [9] also mentioned several Arabian common names for many owl species recorded in Iraq, for example, "*Buha*" for Eagle Owl, "*Khibil*" for Tawny Owl, "*Hama*" for Western Barn Owl, and "*Thabaj*" for Scops owls and likewise.

Moreover, owls in the Arabian culture are symbols of bad omen and embodiment of evil spirits that are roaming in deserts and abandoned places at night to impersonate the wandering people. In their traditional knowledge, the Bedouins of the Iraqi and Syrian deserts are considering owls as flying spirits of dead warriors, who are seeking revenge when they hoot. In other stories, owls are considered as a sign of misfortune, which is horrified and feared by Arabs. They believed that owls bring death when perched on a person's house. Even more, "*follow the owl which leads you to the ruin*", an old Arabian proverb describing owls as a sign of destruction. For decades, owls with their fierce and protruding eyes resembled *satanic* birds; therefore, they are persecuted to be used in sorcery whenever and wherever possible (Al-Sheikhly [39]). Nowadays, superstitions related to owls as being a sign of misfortune and death were refuted and acquitted by the Islamic religion, which highlighted their ecological importance in controlling rodents and the outbreak of zoonotic diseases. However, many communities around the Arabian region are still deemed in the beliefs of the pre-Islamic period, unfortunately (see Conservation issues).

## 3. Conservation issues

Among several wild bird species, owls are presented by local communities of Iraq and Syria for several reasons. In Iraq, the illegal trapping of owls and other birds of prey is commonly and regularly practiced by local communities throughout Iraq [40–42]. Each year, thousands of captured owls of different species are trapped by local hunters and exhibited in the local animal markets in major Iraqi cities. As these birds are forbidden to be hunted in the Islamic religion, the psychological motivation of why these birds are trapped by local hunters has been investigated [39]. The major motives that drive locals to trap birds of prey are the weak implementation of hunting laws and the absence of religious inducement, which drives local communities to pose a further impact on biodiversity through species persecution. Al-Sheikhly [39] analyzed the causes of why local trappers were conducting illegal acts despite their full knowledge that such practices are banned by Islamic religion and Iraqi law. He found that increased levels of poverty are enforcing local communities to apply further pressure on the native biodiversity to compensate for their shortage of livelihood income. Moreover, the local trappers expressed secretive behavior and hesitated to reveal any information on their exact hunting/trapping vicinities and techniques they used, which complicate the validity of the investigation. Despite of what is mentioned above, hunting of wild species with unjustified needs is prohibited in the Islamic faith; however, the psychological insistence of local trappers/hunters to perform such unethical/illegal practices are still far to be known.

Each year, besides other birds of prey, such as *Aquila* eagles, vultures, harriers, and falcons, different owl species are exhibited in the local animal markets of Iraq and Syria to be sold as cage birds or to be used for traditional medicine or sorcery. On one occasion at the local animal market in Baghdad, six owl species along with other different species of migrant raptors were trapped for one day in the arid plateau of Southwestern Iraq. A total of 65 Western Barn (**Figure 1**), 61 Pallid Scops; 72 Short-eared (**Figure 2**), 22 Long-eared (**Figure 3**), 2 Eurasian Eagle, and 3 Pharaoh Eagle owls were observed. They were sold for prices ranging from 10 to 50 US$, depending on their age and state of health. Birds are often presented in poor health with injuries usually caused by aggressive trapping and/or careless handling and many die, unfortunately [14, 39, 40]. As ideologically close communities of Iraq and Syria, it is possible that trappers' behaviors and practices are similar in both countries. However, due to current political unrest, information on birds of prey, including owls trapping in Syria, is extremely scarce and requires dedicated research. In Iraq, the trappers used several large mist nets and Dho-Gazza traps set near desert oases to trap thirsty and exhausted migrant owls and other birds of prey. On a few occasions, trappers used the less effective Bal-chatri traps to trap small-bodied owls like Scops owls.

As elsewhere in Arabia, owls are a symbol of ominous and traditionally are not preferable to be raised inside houses [14]. Nowadays, this belief has dramatically changed in Iraq; young people are looking for adult owls and owlets to be raised as pets. The "attractive look and secretive behavior" of owls are attracting young buyers who showed interest to buy these unique birds. That admire drives local trappers to persist in their illegal quest by trapping more and more owl species on a regular basis. Moreover, owls and their body parts are used as a talismanic recipe in black sorcery,

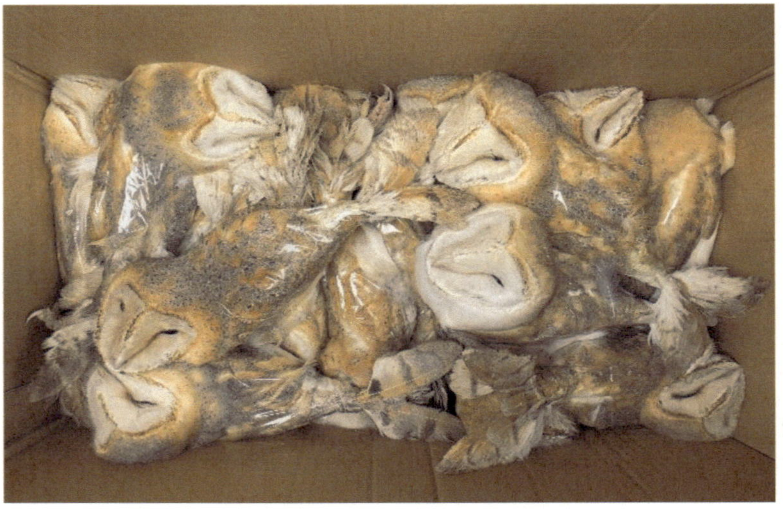

**Figure 1.**
*A large group of migrant Western barn owl Tyto alba cuffed by tape and stacked in a local animal market in Baghdad, Iraq. Photo© Omar Al-Sheikhly.*

**Figure 2.**
*Another group of migrant Short-eared owl Asio flammeus caged in a local animal market in Baghdad, Iraq. Photo© Omar Al-Sheikhly.*

**Figure 3.**
*A group of migrant Long-eared owl Asio otus aggressively caged and presented in a local animal market in Baghdad, Iraq. Photo© Omar Al-Sheikhly.*

which is commonly practiced by some local communities; a desire implies further pressure on owl populations in Iraq and Syria. The view of majority of people's attitude toward owls has not changed since ancient times, as they are a symbol of pessimism and bring bad luck, and are seen by some local residents as the embodiment of the souls of the oppressed dead, and the embodiment of the souls of the old fighters who came for revenge, and many parts of the owl have not stopped so far in the work of magic and sorcery. But recently, some people's view of these birds has changed, and they have turned into ornamental birds that are bred in farms and homes, especially the Western Barn Owl, Eagle Owl, and Tawny Owl, and this is due to the calmness of these birds in general, and they seem to be accustomed to socializing with humans rapidly. Owls have become popular trade. This has led locals to take owl chicks from their nests (without even knowing their species) or catch adults to be offered for sale in local markets or through social sites (**Figure 4**). This act has spread in recent years, especially in light of weak enforcement of hunting laws and trafficking of live animals. According to local wildlife traders' interviewees, some of the large owl species are also smuggled to neighboring countries to be sold as pets, especially in Iraq, Lebanon, and Jordan. Moreover, another factor that indirectly has notably contributed to owls' mortality is due to the use of agricultural pesticides. Large numbers of dead owls were observed in the agricultural lands and cultivated fields, where pesticides were largely used throughout Syria (Ahmad Aidek to Omar Al-Sheikhly pers. comm. 2022).

The weak implementation of strict hunting and wildlife protection laws in Syria has led to the dramatic decline of many wildlife species, including owls. Owls are not only in urgent need of protection, they are part of the Syrian wildlife natural heritage, their ecological necessity in maintaining ecological stability, and key players in the biological control of pests, which warrants conservation. The continuance of such practices combined with other anthropogenic threats (e.g., poisoning, habitat destruction, disturbance ... etc.) could result in a significant decline in the population of resident and migrant owl species in Iraq and Syria. The serious conservation actions

**Figure 4.**
*A juvenile of Pharaoh eagle owl Bubo ascalaphus in the local animal market in Syria. Photo © Ahmad Aidek.*

to address such illegal practices through raising awareness among local communities and enforcing new hunting restrictions are urgently needed to stop such practices, otherwise, such tragedy will continue.

## 4. List of species

### 4.1 Western barn owl *Tyto alba* (Scopoli, 1769) (IUCN red list: LC)

The following are explained in **Figure 5**.

Iraq

Status: Local, but fairly widespread breeding resident, probably regular winter visitor and passage migrant evident by large number of trapped birds during the autumn season (**Figure 1**).

Literature records: Samarra, Mosul, Babylon, between Museyib and Baghdad, Basra, and Fao; nesting between Museyib and Baghdad and Fao [1, 43]; Baghdad [3]; Khanaqin [4, 5, 9], and arid plains of Central Iraq [10].

Recent records: Zurbatiya, Mandli, Mahmmodiya, Desert of Al-Najaf, Kalar, Dalmaj, Falluja, and Haditha.

Syria

Status: Resident breeding. Widespread in all regions except for the dry areas of Al-Badia. But especially along the Euphrates Valley and coastal forests. It inhabits almost

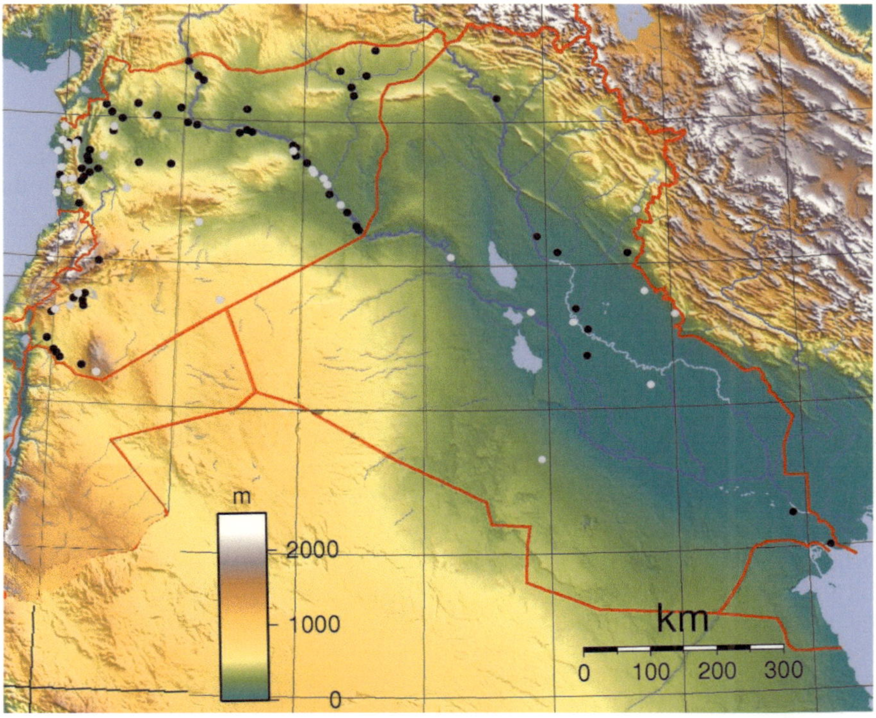


**Figure 5.**
*Distribution of Western barn owl Tyto alba in Iraq and Syria.*

all types of habitats, including farms, trees, deserted buildings, mountain slopes, granaries, caves entrances, and other suitable habitats.

Literature records: Qal'at Al-Hosson [44], Qal'at er-Rahba, Zalabiyyeh, Qal'at Sheizar, Palmyra [45], Qal'at Al-Hosson [46], Dura Europus, Busra, Sabkhat al-Jabbul, Kharabow, Jubb Ramlah, Maskana, Deir Al-Hajar [22], el-Ukersheh, Halabiyyeh, Tell Al-Buderi [47], Tell Mardiekh [48], Yabroud, Qal'at Al-Hosson, Qal'at Salah ad-Din [24], Halabiyyeh, Tell Al-Buderi [49], es-Salihiyyeh, Qal'at er-Rahba, Halabiyyeh, Qal'at Al-Madiq, Tell Al-Buderi, Qal'at Al-Hosson, el-Ukersheh [33], Lake Assad, Dura Europus, Tell Brak [27], Mzeirib, Khrab Al-Shaham, Kharabow [50], Aleppo [51], Yahmool [20], Jub Al-Ghar [52], Busra, es-Salihiyyeh, Qal'at er-Rahba, Zalabiyyeh, Al-Bas'ah, Mari, Al-Marashdeh, Deir Mama, Qal'at Sheizar, Al-Hasaka [28], Ain Jum'aa, Al-Misrab, Qal'at er-Rahba, Qal'at Najim, ez-Zawe [53], Mzeirib, Ain Thakar, Busra, Daraa Dam, Khrab Al-Shaham, es-Salihiyyeh, Ain Jum'aa, Al-Misrab, Qal'at er-Rahba, Zalabiyyeh, Karahta, Kharabow, Sbieneh, Jarablus, Qal'at Najim, Qasr Sallum, Tell Al-Amarneh, Tell Hadia, Yahmool, Al-Sqeilbiyeh, Qal'at Al-Madiq, Qasr Ibn Wardan, Shah Ranaz, Heymu, Tell Beydar, Atheria, Qal'at Al-Hosson, Ebla Ruins, Kafr Daryian, Tell Sandal, Qal'at Salah ad-Din, Jub Al-Ghar, Maseel (Nab'a) Al-Fawar, ar-Raqqa 10 km S of Euphrates, Beer Al-Hashem, el-Ukersheh, Qater Maghara, Qal'at Al-Marqab [17], and Abu Qubays [30].

Recent records: Abu Hardoub, Abu Qubays, Al-Bahlouliyah, Al-Furunlok, Al-Kadmous, Al-Mashqoq, Al-Qallou', Al-Shardoub Forest, at-Tanaf, at-Tebni, Daraa, Deir ez-Zor, Haweijt Saqur, Jaramana, Jnainat Raslan, Khan Sheikhoun, Muhasan, Palmyra, Salamiyah, Sanouber, Saraqeb, Se'lu, Slenfeh, Sreijes, Tartous, and Tell Al-Shaham.

### 4.2 Little owl *Athene noctua* (Scopoli, 1769) (IUCN red list: LC)

The following are explained in **Figure 6**.

Iraq

Status: Fairly widespread breeding resident in North and Central Iraq, uncommon in the south; possibly a passage migrant. Records verified by Salim et al. [15] indicated that all Little Owl observations from Iraq were from Hutton's Owl subspecies *A. n. bactriana* (also see [9]), from Southeastern Azerbaijan, Eastern Iraq, Iran, and Afghanistan east through Central Asia to the Balkhash Lake. However, this range is overlapping with the Lilith Owlet ssp. *A. n. lilith* in Eastern Iraq, which is regarded as having uncertain status; an observation from Southeastern Iraqi arid plains showed features of *A. n. lilith* where it may breed.

Literature records: Samarra-Tekrit area, Amara, Shaiba, Adhaim, Suleimanla [1], Khanaqin [4, 5], Agarguf, Al-Yosifiyah, Baghdad-Habaniyah area, Jabal Himrin, Sir Amadia, Rawandoz [3], and specimen shot in Samarra [9].

Recent records: Haditha-Rutba, Khan Al-Baghdadi, Ramadi, Zurbatiyah, and Teeb (probably of the race *A. n. lilith*).

Syria

Status: Resident breeding. Widespread in all regions. This species can be observed everywhere except for cities and residential areas. It is more common than the Western Barn Owl.

Literature records: Qal'at Sukkara [49], Qal'at Al-Madiq, Shah Ranaz, Qal'at Sukkara, el-Ukersheh, Hirquleh [33], as-Suwar, es-Salihiyyeh, Halabiyyeh, Zalabiyyeh, Mari, Qal'at er-Rahba, Tell Sheikh Hamad, Maalula, Qasr Al-Hayr Al-Sharqi, ar-Rasafah, 10 km E of ar-Rasheidah [32], Qasr Ibn Wardan [28], Mzeirib, Ain

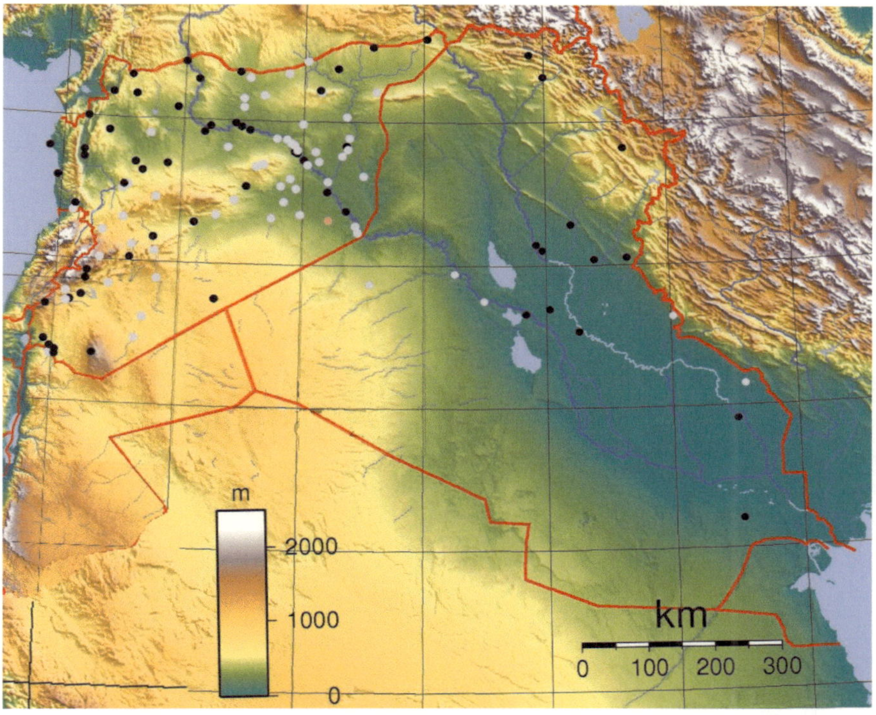


**Figure 6.**
*Distribution of Little owl Athene noctua in Iraq and Syria.*

Thakar, Jelleen, Khrab Al-Shaham, Ain Jum'aa, Mari, Qal'at er-Rahba, Tell Sheikh Hamad, Arnah, Hele, Kharabow, Sbieneh, Babennes, Jarablus, Qal'at Najim, Qal'at Samaan, Qasr Sallum, Yahmool, Qal'at Al-Madiq, Qasr Ibn Wardan, as-Salha, Heymu, Tell Beydar, Atheria, as-Sa'an, at-Tanaf, Al-Qaryatein, Mhassa, Qal'at Al-Hosson, Qal'at Fakhr ad-Din al-Maani, Qasr Al-Hayr Al-Gharbi, Qasr Al-Hayr Al-Sharqi, Tell Senan, Ebla Ruins, Khrab Sultan, Booqa, Fekhaykha, Qal'at Ja'abar, Tell Abiad, Tell Abu Hurera, Ain Arab, Qal'at Al-Marqab [17], ad-Dukhoul [54], and Halabiyyeh [55].

Recent records: Abu ash-shamat, Abu Kamal, ach-Chola, Abu Khashab, ad-Dukhoul, Al- Bolaiyah, Al-Basiri, Al-Bseireh, Al-Furuklus, Al-Ghazili, Al-Hjeifat steppe, Al-Kasra, Al-Nabk, Al-Qaryatein, Al-Qsupy, Al-Thlithawat, Arak, ar-Rasafah, ar-Rawda steppe, ash-Shaddady, as-Sab' Biar, as-Sarayim, as-Sukhnah, as-Suwar, Ayyash, az-Zelif, Beer Al-Hashem, Buhayrat Al-Khatuniyah, Burqan, Feidhat Ibn Muwyin'e, Gleighim, Hasia, Jabal Abd Al-Aziz, Jabal Al-Bilaas, Kabajep, Khadhra Almay, Khanaser, Maadan, Mabroukah, Mari, Marqada, Mueayzilah W of Khabour River, Mueayzilah S of Deir ez-Zor, Mueileh, Qudsayya, Ras Al-Ayn N of Al Hasaka, Sahnaya, Salamiyah, Slouq, Talkalakh, and Twal Al-Aba (**Figures 7–10**).

### 4.3 Eurasian scops owl *Otus scops* (Linnaeus, 1758) (IUCN red list: LC)

The following are explained in **Figure 11**.

Iraq

Status: Fairly widespread breeding summer visitors and passage migrants.

**Figure 7.**
*An adult little owl Athene noctua (ad-Dukhoul) in Syria (race A. n. bactriana). Photo © Ahmad Aidek.*

**Figure 8.**
*An adult little owl Athene noctua near Haditha-Rutba road in extreme Western Iraq (race A. n. bactriana). Photo © Omar Al-Sheikhly.*

**Figure 9.**
*An adult little owl Athene noctua in Najaf Desert of southwestern Iraq (probably of the race A. n. bactriana). Photo © Ali N. Al-Barazangi.*

**Figure 10.**
*An adult little owl Athene noctua in the hills of Teeb in southeastern Iraq (race A. n. lilith). Photo © Omar Al-Sheikhly.*

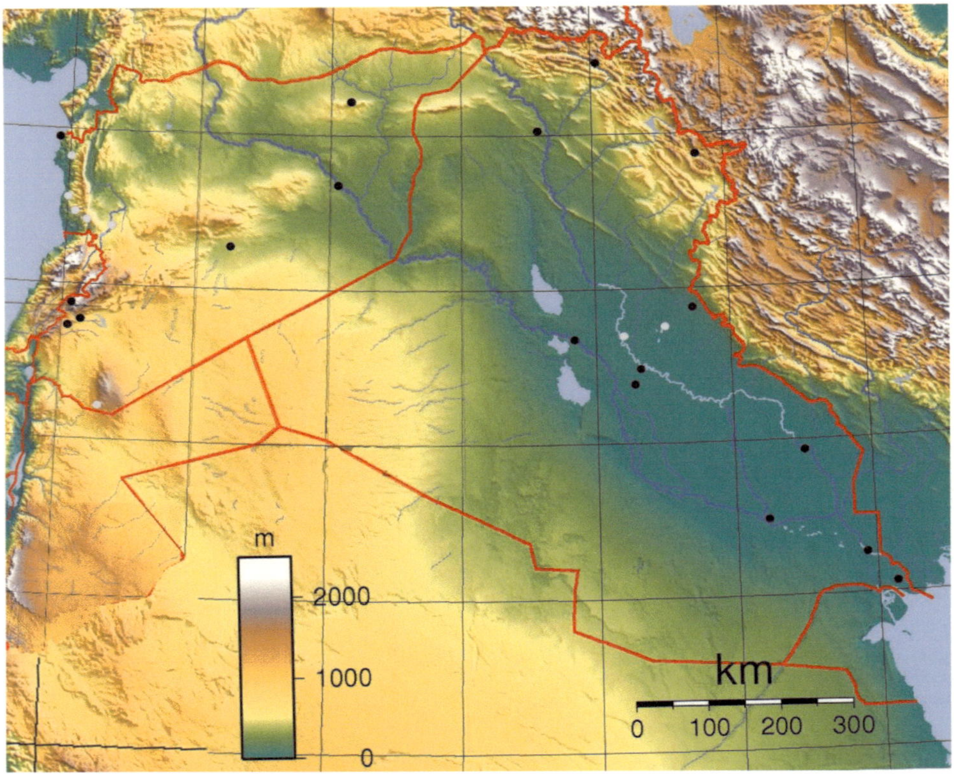


**Figure 11.**
*Distribution of Eurasian scops owl Otus scops in Iraq and Syria.*

Literature records: Basra, Nasirijeh, Museyib, Hilla, Amara, and Fao [1, 43], Habbaniya [56], Khanaqin [4, 5], Barzan and Chowarta [3], and near Mosul [9].

Recent records: Baghdad and Baquba.

Syria

Status: Summer visitor breeding and passage migrant. Spread in all regions except dry areas of Al-Badia. It occurs in small numbers roosting on the trees.

Literature records: Ras Al-Baseet, Qatana [22], Bloudan [57], Talila Reserve [29, 58], Damascus [17], Jabal Abd Al-Aziz [30], and Deir ez-Zor [55].

Recent records: Al-Ghariyeh, Al-Qlatiyeh, aj-Jankeil, Drekeish, Homs, Sreijes (**Figure 12**).

## 4.4 Pallid scops owl *Otus brucei* (Hume, 1873) (IUCN red list: LC)

The following are explained in **Figure 13**.

Iraq

Status: Local resident and/or breeding summer visitor.

Literature records: Kazimain of Baghdad, Hilla and Museyib [1, 9], and University Campus of Baghdad [8].

Recent records: University of Baghdad, Rashidiyah, and Deleasha.

Syria

**Figure 12.**
*An adult Eurasian scops owl Otus scops in Baghdad, Iraq. Photo © Ali N. Al-Barazangi.*

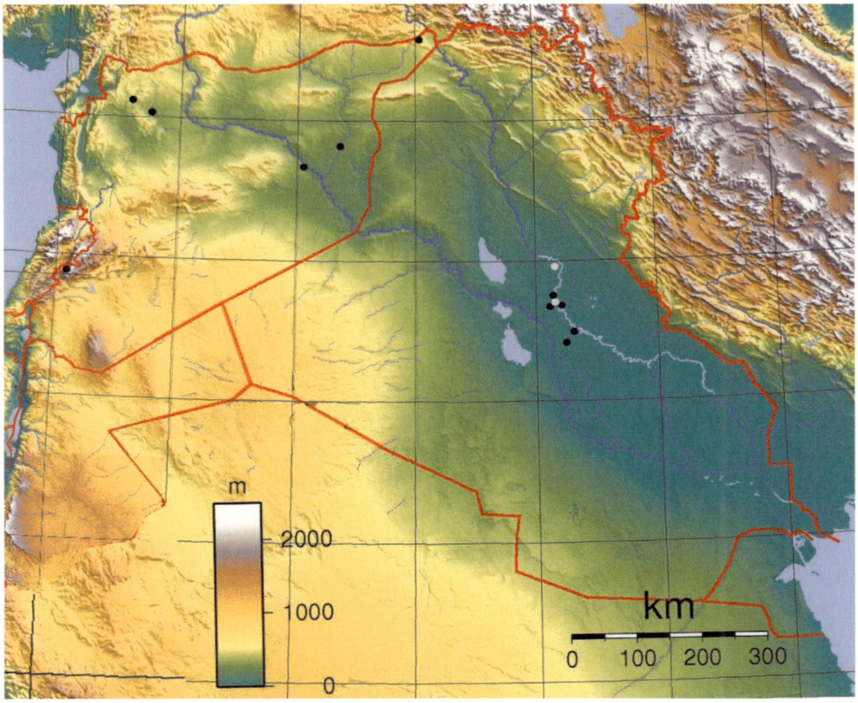


**Figure 13.**
*Distribution of Pallid scops owl Otus brucei in Iraq and Syria.*

**Figure 14.**
*An adult Pallid scops owl Otus brucei in Deleasha in northern Iraq. Photo © Korsh Ararat.*

Status: Rare summer visitor and passage migrant. Spread in all regions except Al-Badia and dry areas.

Literature records: Aleppo [21], Sabkhat al-Jabbul [23], Tell Sheikh Hamad [22], Deir ez-Zor [59], and Serghaya, as-Salha (**Figure 14**) [17].

## 4.5 Long-eared owl *Asio otus* (Linnaeus, 1758) (IUCN red list: LC)

The following are explained in **Figure 15**.

Iraq

Status: Rare breeding residents and winter visitors in northern, numbers of trapped birds observed in autumn suggested heavy passage migrant and probably wintering in low numbers in Central and Southern Iraq (**Figure 3**).

Literature records: Amara, Nasarijeh, Baghdad, and Legait [1], near Great Zab River area [3], Mousl [9], and from Northern Iraq [60].

Recent records: Large numbers of owls trapped somewhere in the southwestern desert of Iraq (see **Figure 3**). University of Baghdad.

Syria

Status: Winter visitor, breeding in Euphrates valley and coast region. Spread in all regions. Mainly found usually in the forested area among pine, cedar, or eucalyptus trees.

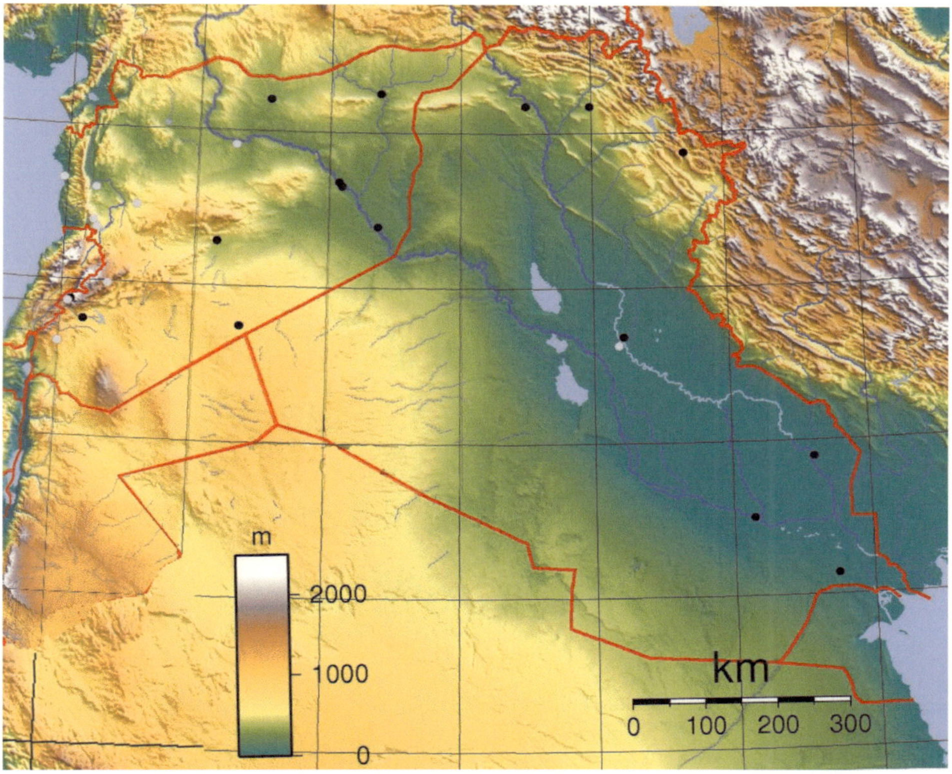


**Figure 15.**
*Distribution of Long-eared owl Asio otus in Iraq and Syria.*

Literature records: Skabro [61], Palmyra [24, 62], Gharaneij [27], Al-Hasaka, Palmyra [28], Damascus, at-Tanaf [17], Bloudan [30], Al-Hseiniyeh [54], and Deir ez-Zor [55],

Recent records: Kherbat Al-Hamam, Jubb Ramlah, Sokas,Al-Koum, Salamiyah, Sheen, Yabroud, az-Zabadani, and Tabqah (**Figures 16** and **17**).

### 4.6 Short-eared owl *Asio flammeus* (Pontoppidan, 1763) (IUCN red list: LC)

The following are explained in **Figure 18**.

Iraq

Status: Uncommon but widespread winter visitors and probably heavy on passage based on large numbers of trapped birds in autumn (**Figure 2**).

Literature records: Fao [43], Amara, Kut, Qalet Saleh, Suleimania [1], Baghdad and Aziziyah [3], and Mosul [9].

Recent records: Only one bird flashed at sunset in the Central Marshes in Southern Iraq.

Syria

Status: Very rare winter visitor. It was recorded only at two sites in the middle of Al-Badia in Central Syria.

Literature records: Palmyra [22] and Talila Reserve [62].

**Figure 16.**
*A migrant adult Long-eared owl Asio otus at the University of Baghdad, Iraq. Photo © Omar Al-Sheikhly.*

**Figure 17.**
*Adult Long-eared owl Asio otus at Al-Hseiniyeh. Photo © Ahmad Aidek.*

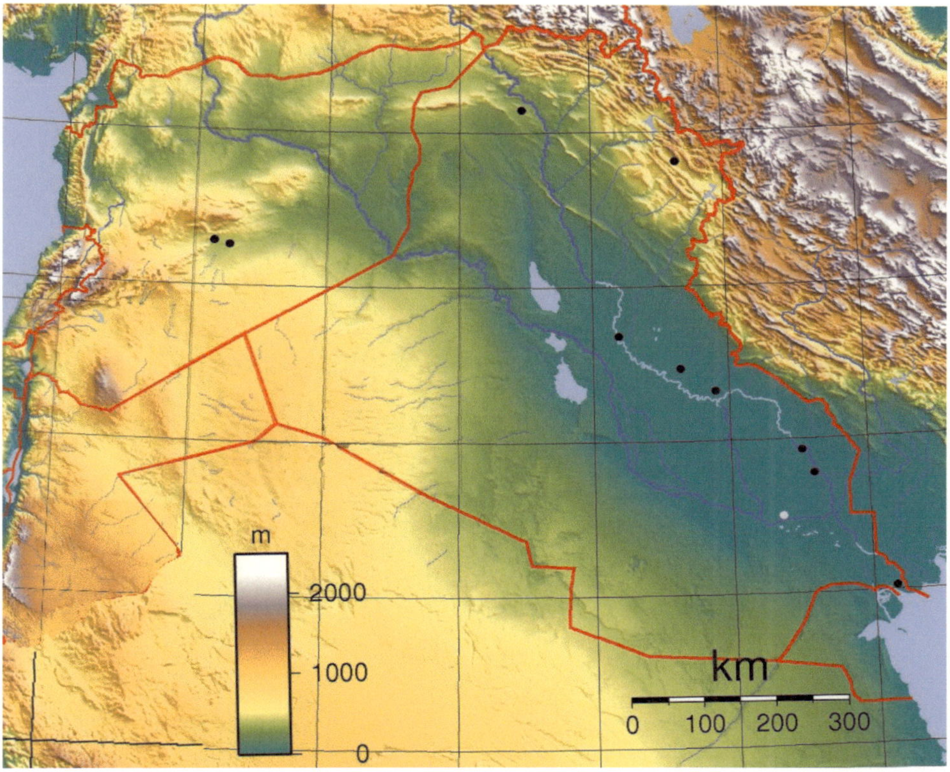


**Figure 18.**
*Distribution of Short-eared owl Asio flammeus in Iraq and Syria.*

### 4.7 Eurasian eagle owl *Bubo bubo* (Linnaeus, 1758) (IUCN red list: LC)

The following are explained in **Figure 19**.

Iraq

Status: Breeding residents mainly in the wooded mountains and rocky hills of northern, found in the elevated grounds of Central, Eastern, and Southern Iraq. Careful identification is required for birds in the rocky hills of Southeastern Iraq as an overlap with Pharaoh Eagle Owl may exist.

Literature records: Jebel Hamrin, Tekrit, ruins of Babylon, Shahroban, Sulemania, Mosul [1], near Khazr River and Aski Kalak [3], Kurkuk, Samarra, and Baghdad [9], and arid steppes of wadi Al-Tharthar in Central Iraq [10].

Recent records: Zraran, Peramagroon Mountain, and QaraDagh Mountain.

Syria

Status: Breeding resident. It occurs in small numbers and is considered rare. It is found in open arid and semi-arid areas devoid of dense trees and often roosts on large rocks or caves. Careful identification is required for birds in the Al-Badia of Central Syria as an overlap with Pharaoh Eagle Owl may exist.

Literature records: 15 km SE of Al-Qaryatein [25], Hamama [26], Dura Europus [27], Palmyra [28, 29], 3 km SW Jelleen, es-Salihiyyeh, Qatana, Qal'at Najim, Mesiaf, Tell Mardiekh (Shehab and Johnson 2009), and Jabal Abd Al-Aziz [30].

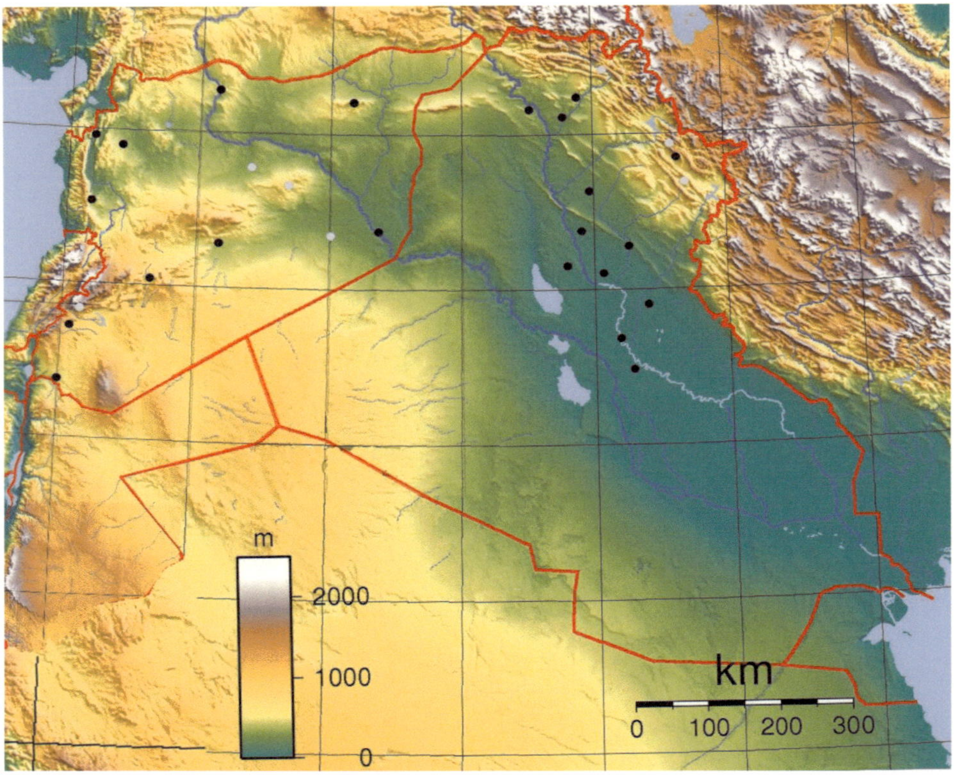


**Figure 19.**
*Distribution of Eurasian eagle owl Bubo bubo in Iraq and Syria.*

Recent records: Jabal Al-Bishri, Halbon, Feidhat Ibn Muwyin'e, and 5 km S ar-Rasafah (**Figure 20**).

### 4.8 Pharaoh eagle owl *Bubo ascalaphus* Savigny, 1809 (IUCN red list: LC)

The following are explained in **Figure 21**.

Iraq

Status: A rare breeding resident in the deserts of Western, Southern, and Southeastern Iraq.

Literature records: Haditha [12, 63], Wadi Al Ubayiadh and Al-Raoudha in Western Iraq [13], Wadi Al-Ga'ara, north of Rutba, ruins of the old city of Ur, Jabal Sanam, Al-Shirhani in Al-Teeb, and Zurbatiyah foothills [14].

Recent records: Desert of Najaf in Southwestern Iraq.

Syria

Status: Rare resident breeding. It was recorded in Al-Badia only.

Literature records: Only one record at Talila Reserve [29].

Recent records: Abu Hbilat (**Figures 22–24**).

### 4.9 Brown fish owl *Ketupa zeylonensis* (Gmelin, 1788) (IUCN red list: LC)

The following are explained in **Figure 25**.

**Figure 20.**
*An adult Eurasian eagle owl Bubo bubo in QaraDagh Mountain in northern Iraq. Photo © Korsh Ararat.*

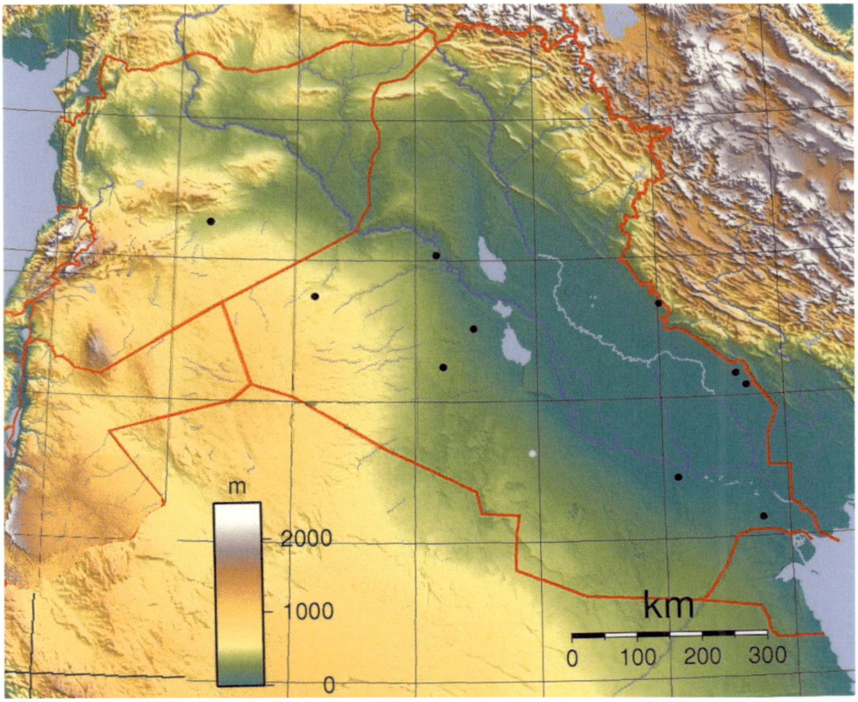


**Figure 21.**
*Distribution of Pharaoh eagle owl Bubo ascalaphus in Iraq and Syria.*

**Figure 22.**
*A breeding adult Pharaoh eagle owl Bubo ascalaphus in extreme Western Iraq. Photo © Omar Al-Sheikhly.*

**Figure 23.**
*An adult Pharaoh eagle owl Bubo ascalaphus in southwestern desert of Iraq. Photo © Ali N. Al-Barazangi.*

**Figure 24.**
*An adult Pharaoh eagle owl Bubo ascalaphus in Talila reserve. Photo © Ahmad Abdullah.*

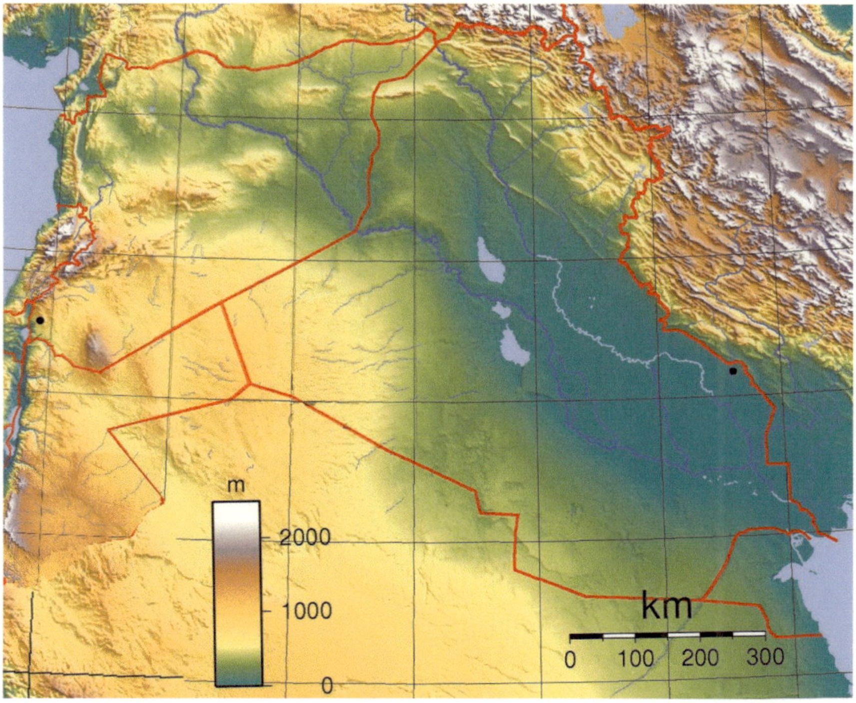


**Figure 25.**
*Distribution of Brown fish owl Ketupa zeylonensis in Iraq and Syria.*

Iraq

Status: Uncertain; probably a very rare resident, not recorded since the 1920s but could still occur near wetlands of Southeastern Iraq.

Literature records: Hills of Eastern Ali Al-Gharbi [2].

Syria

Status: Uncertain, possibly still occurs in Northern Syria.

Literature records: The last record was in the Golan in 1976 [22]. Reports from Northern Syria; further research is needed.

## 4.10 Tawny owl *Strix aluco* Linnaeus, 1758 (IUCN red list: LC)

The following are explained in **Figure 26**.

Iraq

Status: Status uncertain; possibly a rare breeding resident in northern wooded hills as birds heard hooting in February–April; an owl heard in Central Iraq possibly on passage; also, a rare winter visitor.

Literature records: Duhok [2], Habbaniya [56], and Mousl [3, 9].

Recent records: Sir Amadiya and QaraDagh Mountain.

Syria

Status: Breeding resident. Restricted to wooded forests, spread in all regions except in the dry areas of Al-Badia.

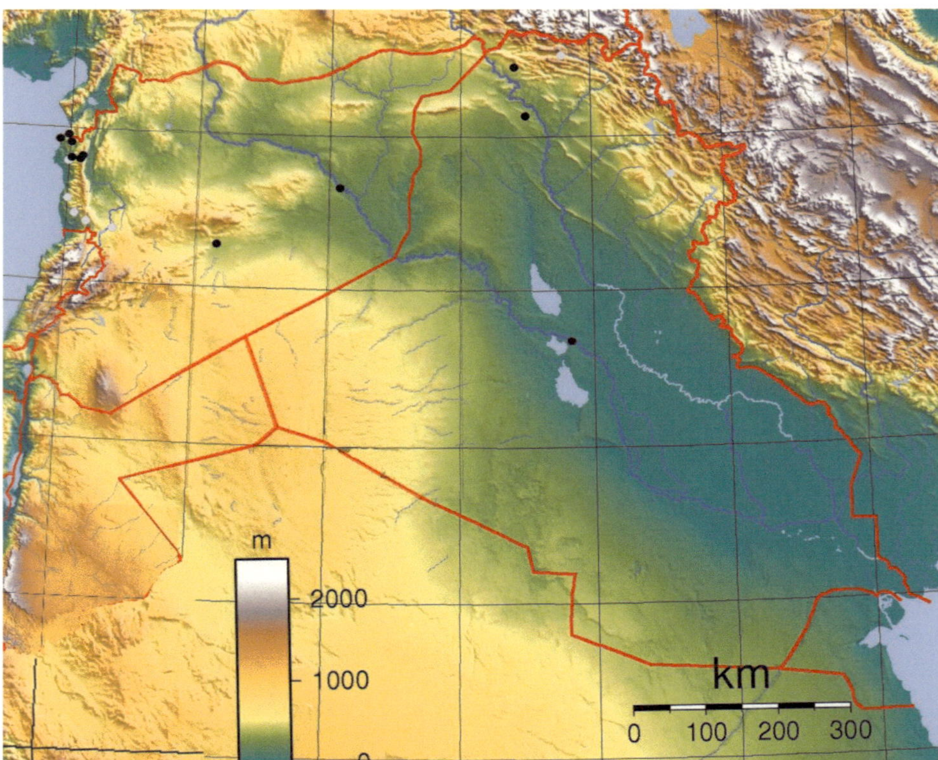


**Figure 26.**
*Distribution of Tawny owl Strix aluco in Iraq and Syria.*

**Figure 27.**
*An adult Tawny owl Strix aluco in QaraDagh Mountain in northern Iraq. Photo© Korsh Ararat.*

Literature records: Jubb Al-Ghar [17], Kasab, Ras Al-Baseet, Slenfeh [22], Palmyra [29], ar-Rawda near Rabi'ah, Qal'at Salah ad-Din [30], and Deir ez-Zor, Palmyra [64].

Recent records: Drekeish, Homs, Sreijes, Al-Qlatiyeh, and aj-Jankeil (**Figure 27**).

## Acknowledgements

We would like to express our gratitude to Miss. Yesra Shkaky (General Organization of Remote Sensing, Department of Environmental Researches and Regional Planning, Syria) for her kind technical support in the preparation of the species distribution maps. We are grateful to Dr. Mukhtar K. Haba, Dr. Nadheer A. Fazaa (University of Baghdad-Iraq), Korsh Ararat (University of Sulaymani-Iraq), and Dr. Zuhair S. Amr (Jordan) for their valuable comments on owls of Iraq and Syria along with their valuable remarks on the initial draft of this chapter. Our thanks extend once more time to Korsh Ararat and Ali N. Al-Barazangi (Iraqi Ministry of Environment-Iraq) and to Ahmad Abdullah (Conservationist-Syria) for providing owls' photographs featured in this chapter. In appreciation of the impressive work that had been contributed to add knowledge to the owl avifauna of Syria by the late Dr. Adwan Shehab, we dedicate this chapter to him, wishing mercy and forgiveness for his soul.

# Appendix

## Annex I. Site gazetteer in Iraq

| Site Name | Coordinates |
|---|---|
| Zurbatiya (Zurbatiyah) | 33°17′30.00"N 46° 6′5.66″E |
| Mandli | 33°49′55.23"N 45°32′40.78″E |
| Mahmmodiya | 33° 1′28.99"N 44°23′22.69″E |
| Desert of Al-Najaf | 30°58′55.48"N 43°59′53.13″E |
| Kalar | 34°36′13.69"N 45°23′45.33″E |
| Dalmaj | 32°14′54.81"N 45°33′6.72″E |
| Falluja | 33°16′33.70"N 43°47′59.89″E |
| Haditha | 34° 1′39.98"N 42°21′21.61″E |
| Haditha-Rutba | 33°29′13.24"N 41° 5′48.18″E |
| Khan Al-Baghdadi | 33°52′54.24"N 42°33′12.98″E |
| Ramadi | 33°28′16.41"N 43°15′58.71″E |
| Teeb | 32°21′5.64"N 47°23′58.03″E |
| Baghdad | 33°28′16.96"N 44°18′26.56″E |
| Baquba | 33°46′33.09"N 44°38′54.73″E |
| University of Baghdad | 33°16′32.57"N 44°22′41.12″E |
| Rashidiyah | 33°30′42.86"N 44°19′46.65″E |
| Deleasha | 35°49′11.88"N 45°27′15.74″E |
| Central Marshes | 31° 3′47.57"N 47° 5′52.92″E |
| Zraran | 37° 6′17.60"N 44° 0′44.09″E |
| Peramagroon Mountain | 35°45′31.65"N 45°13′41.49″E |
| QaraDagh Mountain | 35°14′20.25"N 45°22′21.87″E |
| Sir Amadiya | 37° 4′53.60"N 43°29′51.48″E |

## Annex II. Site gazetteer in Syria

| **10 km S of Euphrates** | **35°50′43.0"N 38°54′50.0″E** |
|---|---|
| 15 km SE of Al-Qaryatein | 34°05′53.0"N 37°17′02.0″E |
| 3 km SW Jelleen | 32°44′17.0"N 35°57′01.0″E |
| 5 km S ar-Rasafah | 35°33′53.8"N 38°46′26.0″E |
| Abu ash-shamat | 33°40′02.1"N 36°53′40.5″E |
| Abu Hardoub | 34°51′10.6"N 40°37′14.3″E |
| Abu Hbilat | 35°01′51.7"N 37°18′50.5″E |
| Abu Kamal | 34°27′18.1"N 40°55′45.5″E |

| **10 km S of Euphrates** | **35°50′43.0"N 38°54′50.0″E** |
|---|---|
| Abu Khashab | 35°56′46.9"N 40°00′47.1″E |
| Abu Qubays | 35°14′43.1"N 36°18′36.5″E |
| Ach-Chola | 35°11′05.2"N 39°48′16.5″E |
| Ad-Dukhoul | 35°02′54.0"N 39°54′29.6″E |
| Ain Jum'aa | 35°26′04.0"N 40°02′41.2″E |
| Ain Thakar | 32°51′34.1"N 35°54′13.8″E |
| Aj-Jankeil | 35°35′44.0"N 36°01′56.8″E |
| Al- Bolaiyah | 35°22′59.9"N 39°20′59.6″E |
| Al-Bahlouliyah | 35°38′15.0"N 35°57′23.7″E |
| Al-Bas'ah | 35°40′19.6"N 39°49′38.2″E |
| Al-Basiri | 34°09′18.1"N 37°36′35.0″E |
| Al-Bseireh | 35°09′20.9"N 40°25′53.8″E |
| Aleppo | 36°12′15.4"N 37°10′44.9″E |
| Al-Furuklus | 34°36′10.3"N 37°05′07.7″E |
| Al-Furunlok | 35°49′50.3"N 35°59′36.6″E |
| Al-Ghariyeh | 32°23′43.1"N 36°38′59.4″E |
| Al-Ghazili | 36°18′00.1"N 39°00′28.5″E |
| Al-Hasaka | 36°30′34.2"N 40°45′57.7″E |
| Al-Hjeifat steppe | 35°34′29.6"N 40°13′32.0″E |
| Al-Hseiniyeh | 35°21′27.1"N 40°08′50.0″E |
| Al-Kadmous | 35°06′13.1"N 36°09′40.7″E |
| Al-Kasra | 35°33′56.1"N 39°55′00.9″E |
| Al-Koum | 33°12′22.6"N 35°57′33.5″E |
| Al-Marashdeh | 34°30′25.4"N 40°55′21.1″E |
| Al-Mashqoq | 32°25′32.9"N 36°43′39.6″E |
| Al-Misrab | 35°32′45.5"N 39°51′21.3″E |
| Al-Nabk | 34°01′13.2"N36°43′45.2″E |
| Al-Qallou' | 35°15′07.9"N 35°57′04.9″E |
| Al-Qaryatein | 34°13′45.9"N 37°14′23.0″E |
| Al-Qlatiyeh | 34°47′50.5"N 36°19′02.0″E |
| Al-Qsupy | 35°44′14.7"N 39°45′55.1″E |
| Al-Shardoub Forest | 35°36′01.3"N 36°03′35.1″E |
| Al-Sqeilbiyeh | 35°22′12.0"N 36°23′58.0″E |
| Al-Thlithawat | 35°21′19.8"N 39°14′57.4″E |
| Arak | 34°40′09.6"N 38°35′46.8″E |
| ar-Rasafah | 35°37′16.3"N 38°46′08.9″E |
| ar-Rawda near Rabi'ah | 35°48′54.2"N 36°01′47.6″E |
| ar-Rawda steppe | 35°14′19.8"N 41°01′07.8″E |

| | |
|---|---|
| **10 km S of Euphrates** | **35°50′43.0"N 38°54′50.0″E** |
| ash-Shaddady | 36°03′34.8"N 40°44′06.0″E |
| as-Sab' Biar | 33°46′00.1"N 37°40′28.5″E |
| as-Salha | 37°09′44.0"N 42°04′19.0″E |
| as-Sarayim | 34°38′00.1"N 39°30′28.5″E |
| as-Sukhnah | 34°53′01.6"N 38°53′05.8″E |
| as-Suwar | 35°30′37.7"N 40°39′50.5″E |
| Atheria | 35°22′09.9"N 37°46′36.1″E |
| at-Tanaf | 33°29′24.1"N 38°39′52.6″E |
| at-Tebni | 35°36′14.5"N 39°49′28.1″E |
| Ayyash | 35°25′19.3"N 40°03′35.7″E |
| az-Zabadani | 33°43′26.0"N 36°06′05.2″E |
| az-Zelif | 32°55′30.1"N 37°20′15.4″E |
| Beer Al-Hashem | 36°10′33.0"N 39°01′36.6″E |
| Bloudan | 33°43′29.2"N 36°07′37.8″E |
| Buhayrat Al-Khatuniyah | 36°24′38.2"N 41°13′41.8″E |
| Burqan | 35°02′46.7"N 37°07′50.9″E |
| Busra | 32°31′07.7"N 36°28′54.7″E |
| Damascus | 33°30′47.8"N 36°16′52.3″E |
| Daraa | 33°19′32.0"N 36°14′36.5″E |
| Daraa Dam | 32°36′04.0"N 36°06′56.9″E |
| Deir Al-Hajar | 33°21′06.1"N 36°26′59.0″E |
| Deir ez-Zor | 35°20′21.0"N 40°08′31.3″E |
| Deir Mama | 35°08′30.0"N 36°19′53.5″E |
| Drekeish | 34°53′44.9"N 36°08′07.2″E |
| Ebla Ruins | 35°47′54.0"N 36°47′52.0″E |
| el-Ukersheh | 35°51′39.5"N 39°07′12.5″E |
| es-Salihiyyeh (Dura Europus) | 34°44′55.5"N 40°43′47.7″E |
| ez-Zawe | 35°04′55.0"N 36°18′11.0″E |
| Feidhat Ibn Muwyin'e | 34°41′33.8"N 39°59′02.7″E |
| Gharaneij | 34°47′04.6"N 40°43′07.9″E |
| Gleighim | 33°13′31.0"N 37°28′33.1″E |
| Golan Heights | 33°00′04.7"N 35°44′19.5″E |
| Halabiyyeh | 35°39′57.3"N 39°49′49.1″E |
| Halbon | 33°39′51.8"N 36°14′57.2″E |
| Hamama | 35°55′26.7"N 36°21′24.0″E |
| Hasia | 34°24′36.8"N 36°45′43.9″E |
| Haweijt Saqur | 35°18′57.7"N 40°10′31.6″E |
| Heymu | 37°02′31.0"N 41°09′45.0″E |

| | |
|---|---|
| **10 km S of Euphrates** | **35°50′43.0"N 38°54′50.0″E** |
| Homs | 34°43′40.1"N 36°43′10.0″E |
| Jabal Abd Al-Aziz | 36°25′13.7"N 40°19′15.4″E |
| Jabal Al-Bilaas | 34°53′46.9"N 37°36′57.7″E |
| Jabal Al-Bishri | 35°20′00.1"N 39°20′31.5″E |
| Jarablus | 36°49′09.4"N 38°00′54.5″E |
| Jaramana | 33°29′18.0"N 36°20′41.1″E |
| Jnainat Raslan | 34°55′51.5"N 36°07′19.3″E |
| Jub Al-Ghar | 35°38′15.9"N 36°13′29.7″E |
| Jubb Ramlah | 35°12′29.4"N 36°25′56.8″E |
| Kabajep | 35°04′43.4"N 39°39′44.2″E |
| Kafr Daryian | 36°10′26.3"N 36°39′57.9″E |
| Karahta | 33°24′33.1"N 36°25′36.6″E |
| Kasab | 35°55′35.3"N 35°59′17.0″E |
| Khadhra Almay | 34°52′02.4"N 39°43′00.4″E |
| Khan Sheikhoun | 35°26′32.4"N 36°39′09.0″E |
| Khanaser | 35°46′58.9"N 37°29′33.1″E |
| Kharabow | 33°30′19.8"N 36°27′55.0″E |
| Kherbat Al-Hamam | 34°41′47.0"N 36°28′58.9″E |
| Khrab Al-Shaham | 32°39′38.0"N 36°01′45.0″E |
| Lake Assad | 35°56′26.7"N 38°11′13.6″E |
| Maadan | 35°45′09.8"N 39°35′28.7″E |
| Mabroukah | 36°39′16.5"N 39°45′39.5″E |
| Mari | 34°33′01.6"N 40°53′22.8″E |
| Marqada | 35°45′25.5"N 40°46′07.8″E |
| Maseel (Nab'a) Al-Fawar | 33°13′40.0"N 35°56′51.0″E |
| Maskana | 35°57′55.1"N 38°02′39.8″E |
| Mesiaf | 35°03′55.7"N 36°20′57.5″E |
| Mueayzilah | 34°37′14.6"N 40°26′44.5″E |
| Mueileh | 35°37′15.0"N 40°44′30.5″E |
| Muhasan | 35°13′15.1"N 40°20′14.3″E |
| Mzeirib | 32°42′11.4"N 36°01′49.3″E |
| Palmyra | 34°34′30.1"N 38°17′45.3″E |
| Qal'at Al-Hosson | 34°45′25.4"N 36°17′42.3″E |
| Qal'at Al-Madiq | 35°25′10.6"N 36°23′31.2″E |
| Qal'at Al-Marqab | 35°09′03.0"N 35°56′55.0″E |
| Qal'at er-Rahba | 35°00′17.8"N 40°25′24.2″E |
| Qal'at Najim | 36°33′18.8"N 38°15′43.2″E |
| Qal'at Salah ad-Din | 35°35′43.7"N 36°03′22.2″E |

| | |
|---|---|
| **10 km S of Euphrates** | **35°50′43.0”N 38°54′50.0″E** |
| Qal’at Sheizar | 35°15′57.3”N 36°33′53.4″E |
| Qasr Ibn Wardan | 35°22′23.0”N 37°15′14.6″E |
| Qasr Sallum | 36°10′01.3”N 37°55′17.3″E |
| Qatana | 33°26′15.4”N 36°06′01.4″E |
| Qater Maghara | 35°53′06.5”N 39°01′27.2″E |
| Qudsayya | 33°31′42.8”N 36°10′56.7″E |
| Ras Al-Ayn | 36°50′42.5”N 40°04′50.5″E |
| Ras Al-Baseet | 35°50′45.8”N 35°50′25.0″E |
| Sabkhat al-Jabbul | 36°02′40.4”N 37°31′01.2″E |
| Sahnaya | 33°25′34.1”N 36°13′10.4″E |
| Salamiyah | 35°00′33.9”N 37°03′44.4″E |
| Sanouber | 35°28′48.3”N 35°53′23.5″E |
| Saraqeb | 35°51′43.8”N 36°48′15.8″E |
| Sbieneh | 33°26′01.0”N 36°17′03.0″E |
| Se’lu | 35°09′09.5”N 40°22′53.0″E |
| Serghaya | 33°48′43.7”N 36°09′31.5″E |
| Shah Ranaz | 35°30′58.3”N 36°24′15.0″E |
| Sheen | 34°46′53.1”N 36°25′17.4″E |
| Skabro | 36°25′08.7”N 39°04′04.5″E |
| Slenfeh | 35°35′54.8”N 36°11′00.3″E |
| Slouq | 36°35′46.6”N 39°07′20.8″E |
| Sokas | 35°18′21.2”N 35°55′53.2″E |
| Sreijes | 35°05′04.0”N 36°00′41.7″E |
| Tabqah | 35°49′53.9”N 38°33′07.2″E |
| Talila Reserve | 34°31′34.2”N 38°31′41.5″E |
| Talkalakh | 34°40′10.6”N 36°15′30.7″E |
| Tartous | 34°53′20.9”N 35°53′50.5″E |
| Tell Abiad | 36°41′32.7”N 38°56′55.7″E |
| Tell Abu Hurera | 35°50′34.7”N 38°23′06.4″E |
| Tell Al-Amarneh | 36°37′01.0”N 38°11′01.0″E |
| Tell Al-Buderi | 36°23′15.2”N 40°48′49.8″E |
| Tell Al-Shaham | 33°14′57.0”N 35°59′42.3″E |
| Tell Beydar | 36°44′11.0”N 40°35′01.9″E |
| Tell Brak | 36°40′35.8”N 41°01′51.9″E |
| Tell Hadia | 35°59′25.7”N 36°56′20.3″E |
| Tell Mardiekh | 35°48′35.6”N 36°47′10.5″E |
| Tell Sandal | 36°03′54.0”N 36°44′48.0″E |
| Tell Senan | 35°03′19.0”N 37°04′33.0″E |

| | |
|---|---|
| **10 km S of Euphrates** | **35°50′43.0"N 38°54′50.0″E** |
| Tell Sheikh Hamad | 35°38′31.0"N 40°44′23.7″E |
| Twal Al-Aba | 36°20′25.1"N 39°21′35.2″E |
| Yabroud | 33°58′23.4"N 36°40′04.4″E |
| Yahmool | 36°34′54.1"N37°07′51.6″E |
| Zalabiyyeh | 35°39′07.6"N 39°51′02.5″E |

## Author details

Omar F. Al-Sheikhly[1*] and Ahmad E. Aidek[2]

1 Department of Biology, College of Science, University of Baghdad, Baghdad, Iraq

2 General Commission for Scientific Agricultural Research, Deir ez-Zor, Syria

*Address all correspondence to: alsheikhlyomar@gmail.com

## References


[1] Ticehurst CB, Buxton PA, Cheesman RE. The birds of Mesopotamia. Journal of the Bombay Natural History Society. 1922;**28**(210–250), 381–427, 650–674,:937-956

[2] Ticehurst CB, Cox P, Cheesman RE. Additional notes on the avifauna of Iraq. Journal of the Bombay Natural History Society. 1926;**31**:91-119

[3] Moore HJ, Boswell C. Field Observations on the Birds of Iraq. Parts I and II 1956, Part III 1957. Baghdad: Iraq Natural History Museum; 1956

[4] Sage BL. Field notes on autumn migration in the Khanaqin area in 1958. Iraq Natural History Museum Publication. 1958;**16**:33-48

[5] Sage BL. Field notes on some birds of eastern Iraq. Ardea. 1960;**48**:160-178

[6] Marchant S. Iraq bird notes - 1960. Bulletin Iraq Natural History Museum. 1961;**1**(4):1-37

[7] Marchant S. Iraq bird notes - 1961. Bulletin Iraq Natural History Museum. 1962;**2**(1):1-40

[8] Scott DA, Carp E. A midwinter survey of wetlands in Mesopotamia, Iraq: 1979. Sandgrouse. 1982;**4**:6-76

[9] Allouse B. [Birds of Iraq]. 1960, 1961, 1962; Vols. I, II, III. Baghdad: Al-Rabita Press; 1960 [In Arabic]

[10] Al-Dabbagh KY. The birds of semi-desert areas of Central Iraq. Sandgrouse. 1998;**20**:135-141

[11] Porter R, Aspinall S, Gale J, Langman M, Small B. Birds of the Middle East. London: Christopher Helm; 2010 384pp

[12] Vaurie C. Systematic Notes on Palearctic Birds. No. 41 Strigidae: The Genus Bubo. American Museum Novitates 1960. The American Museum of Natural History; 2000. pp. 1-31

[13] Al-Sheikhly OF. Breeding of pharaoh eagle owl *Bubo ascalaphus* in Iraq. Sandgrouse. 2012;**34**:72-74

[14] Al-Sheikhly OF, Mikkola H, Mousavi SB. Pharaoh eagle-owl *Bubo ascalaphus* (Savigny 1809) (Stringiformes, Strigidae), the "shrouded in mystery" owl of Iraq and Iran. Bulletin of Iraq Natural History Museum. 2020;**16**(2):219-230

[15] Salim MA, Al-Sheikhly OF, Majeed KA, Porter RF. Annotated checklist of the birds of Iraq. Sandgrouse. 2012;**34**(1):3-44

[16] OSME. OSME Region List of Bird Taxa Bird List. Version 8.1. The Ornithological Society of the Middle East, the Caucasus and Central Asia; 2022. Available at: https://osme.org/orl/ [Accessed on 22 December 2022]

[17] Shehab AH, Johnson DH. Distribution of owls in Syria. In: Johnson DH, Van Nieuwenhuyse D, Duncan JR, editors. Proceeding Fourth World Owl Conference. Oct–Nov 2007, Vol. 97. No. 4. Groningen, The Netherlands: Ardea; 2009. pp. 503–514. Available form: https://www.researchgate.net/publication/260828009_Shehab_A_DH_Johnson_2009_Observational_records_and_distribution_of_seven_owls_in_Syria_Proceedings_from_the_4th_World_Owl_Conference_Groningen_Netherlands_Ardea_974503-514

[18] Raczynski J, Ruprecht AL. The effect of digestion on the osteological

composition of owl pellets. Acta Ornithologica. 1974;**14**:25-38

[19] Andrews P. Owls, Caves and Fossils. Chicago: University of Chicago Press; 1990 231pp

[20] Shehab AH, Al-Charabi S. Food of the barn owl *Tyto alba* in the Yahmool area, northern Syria. Turkish Journal of Zoology. 2006;**30**:175-179

[21] Clarke GVH. Some notes on birds found breeding in the neighborhood of Aleppo in 1919. Ibis. 1924;**6**:101-110

[22] Baumgart W, Kasparek M, Stephan B. The Birds of Syria: An Overview. Heidelberg: Max Kasparek Verlag; 1995 127p

[23] Evans MI. Important Bird Areas in the Middle East. Cambridge: Bird Life International; 1994 410 p

[24] Obuch J. Dormice in the diet of owls in the Middle East. Trakya University Journal of Scientific Research. 2001;**2**: 145-150

[25] Shehab AH. Diet of the eagle owl, *Bubo bubo*, in Syria. Zoology in the Middle East. 2004;**33**:21-26

[26] Shehab AH, Mamkhair IH. First record of the Egyptian fruit bat, *Rousettus aegyptiacus*, from Syria. Zoology in the Middle East. 2004;**33**: 73-78

[27] Murdoch DA, Vos R, Abdallah A, Abdallah M, Andrews I, Al-Asaad A, et al. Final Report of the Syrian Wetland Expedition, January–February 2004. London: A winter survey of Syrian wetlands; 2005 137 p

[28] Benda P, Andreas M, Kock D, Lučan R, Munclinger P, Nová P, et al. Bats (Mammalia: Chiroptera) of the eastern Mediterranean. Part 4. Bat fauna of Syria: Distribution, systematics, ecology. Acta Societatis Zoologica Bohemicae. 2006;**70**:1-329

[29] Serra G, Mirreh M, Kaddour H, Razzouk T, Al-Jundi A, Kanani A, et al. Assessment and Characterization of Al Talila Reserve and Surrounding Palmyrean Desert. IUCN/DGCS Project Report. Syria: IUCN publication; 2009 134 p

[30] Bowler J. Spring trip to Syria 30 March-26 April 2010. Available from: https://osme.org/trip-reports/syria10/ [Accessed: October 10, 2022].

[31] Porter RF, Aspinall S. Birds of the Middle East. In: Princeton Field Guides. Vol. 75. Princeton: Princeton University Press; 2012, 2012 384p

[32] Obuch J, Kristin A. Prey composition of the little owl *Athene noctua* in an arid zone (Egypt, Syria, Iran). Folia Zoologica. 2004;**53**:65-79

[33] Shehab AH, Daoud A, Kock D, Amr Z. Small mammals recovered from owl pellets from Syria (Mammalia: Soricidae, Chiroptera, Rodentia). Zoology in the Middle East. 2004;**33**: 27-42

[34] Murdoch DA, Betton KF. A checklist of the birds of Syria. Sandgrouse. 2008;**2**: 1-48

[35] Mikkola H. Owls of the World Enhanced e-Book. London: Bloomsbury/ Christopher Helm; 2014 528p

[36] Manners G, Diekmann J. Long-eared owl *Asio otus* breeding in north-West Syria. Sandgrouse. 1996;**18**(2):62

[37] Abdulridha S, Hamera Z. Simpole owl in Arabic and Persian culture.

Journal of Literature. 2018;**127**:599-617 [in Persian]

[38] Hussain M, Akhtar J. Owls in Islam and Pakistan: Loopholes and suggestions for conservation. Journal of Bioresource Management. 2021;**8**(1):40-45

[39] Al-Sheikhly OF. Why Are Owls Being Trapped in Iraq? E-Tyto: The International Owl Society; 2021. pp. 7-14

[40] Al-Sheikhly OF. A survey report on the raptors trapping and trade in Iraq. Wildlife Middle East. 2011;**6**(1):6

[41] Al-Sheikhly OF, Al-Barazangi AN, Mukhtar KH, Fazaa N, Abdulzahra HK, Abou Turab MK, et al. Ring recoveries from steppe eagles and eastern Imperial eagles from the Russian and Kazakhstan breeding populations and a review of major threats to eagles in Iraq. Raptors Conservation. 2017;**35**:51-61

[42] Al-Sheikhly OF, Al-Azawi AJ. The diurnal birds of prey (raptors) in the Mesopotamian marshes of southern Iraq with notes on their conservation status. Bulletin of the Iraqi Natural History Museum. 2019;**15**(4): 381-402

[43] Cumming WD. Natural history notes from Fao. Journal of the Bombay Natural History Society. 1918;**26**:292-295

[44] Pradel A. Biometrical remarks on the hamster *Cricetulus migratorius* (Pallas 1773) (Rodentia, mammals) from Krak des chevaliers (Syria). Acta Zoologica. Cracoviensia. 1983;**25**:271-292

[45] Kock D, Nader IA. Pygmy shrew and rodents from the near east (Mammalia: Soricidae, Rodentia). Senckenbergiana biologia. 1981;**64**:13-23

[46] Nadachowski A, Smielowski J, Rzebik-Kowalska B, Daoud A. Mammals from the near east in polish collections. Acta Zoologica Cracoviensia. 1990;**33**(6): 91-120

[47] Kock D. The gerbils and jirds of Syria (Mammalia: Rodentia: Muridae: Gerbillinae). Senckenbergiana Biologica. 1998;**77**:117-122

[48] Shehab AH, Kowalski K, Daoud A. Biometrical remarks on the golden hamster *Mesocricetus auratus* (Waterhouse, 1839) (Cricetidae, Rodentia) from Ebla (northern Syria). Acta Zoologica Cracoviensia. 1999;**42**: 403-406

[49] Hutterer R, Kock D. Recent and ancient records of shrews from Syria, with notes on *Crocidura katinka* bate, 1937 (Mammalia: Soricidae). Bonner Zoologische Beiträge. 2002;**50**: 249-258

[50] Shehab AH. Food of the barn owl *Tyto alba* in southern Syria. Acta Zoologica Cracoviensia. 2005;**48**:35-42

[51] Dubois P. Syrian trip report. 2006. Available from: https://osme.org/trip-reports/syria9/ [Accessed: October 10, 2022].

[52] Shehab A, Mamkhair I, Amr Z. Remains of the water vole, *Arvicola terrestris* (Linnaeus, 1758) (Microtinae, Rodentia), from North–Western Syria. Zoology in the Middle East. 2006;**37**(1): 111-113

[53] Shehab A, Karatas A, Amr Z, Mamkhair I, Sozen M. The distribution of bats (Mammalia: Chiroptera) in Syria. Vertebrate Zoology. 2007;**57**(1):103-132

[54] Aidek A. Guide to the Biodiversity of Deir Ez–Zor Area. Damascus, Syria: Al–Quds House for Sciences; 2010 112 pp. [in Arabic]

[55] Murdoch DA, Aidek A. Birding sites of the OSME region 8–the birds of the lower Syrian Euphrates. Sandgrouse. 2012;**34**:152-176

[56] Chapman EA, McGeoch JA. Recent field observations from Iraq. Ibis. 1956; **98**:577-594

[57] Hofland R, Keijl G. Syrian Sociable Lapwing Survey, 18 February - 5 March 2007. Beek-Ubbergen, Netherlands: WIWO-report 85; 2008 98 pp

[58] Haraldsson T. Trip report: Syria 14–22th April 2009. Available from: https://osme.org/wp-content/uploads/2019/10/Syria_April_2009.pdf [Accessed: October 10, 2022]

[59] Wester J. Syria and Jordan trip report. 1998. Available from: https://osme.org/trip-reports/syria1/ [Accessed: October 10, 2022].

[60] Ararat K, Al-Sheikhly OF, Porter RF, Salim M. Breeding birds in Iraq: Important new discoveries. Sandgrouse. 2011;**33**:12-33

[61] Eccles S. Report on a trip to Syria, March, 2001. Unpublished Report.

[62] Serra G, Al-Qaim G, Abdallah MS, Kanani A, Al-Asaad AK. A long- term bird survey in the Central Syrian desert (2000–2004) part 2-a provisional annotated checklist. Sandgrouse. 2005; **27**:104-125

[63] Balmer D, Harrison I. Around the region. Sandgrouse. 2011;**33**:202

[64] Kinzelbach R. The white-cheeked bulbul, *Pycnonotus leucogenys*, a resident of the Palmyra oasis, Syria. Zoology in the Middle East. 1986;**1**:73-74

Section 5

# Eating Owls as Medicine or for Magic and Witchcraft

## Chapter 7

# Owls Used as Food and Medicine and for Witchcraft in Africa

*Heimo Mikkola*


## Abstract

Wildlife has been used throughout the world since ancient cultures as food or medicine as well as heralds of events and in magic or witchcraft activities. Owl belief interview studies were undertaken in 20 African countries between 1996 and 2002. A total of 794 interviewed people reported reasons for 333 owl killings. In 17 percent of the cases, owls were killed because they represent an omen of death or disaster. In 16 percent of cases, owls were killed for food. Particularly in war-stricken countries, owls are often eaten, like in Sierra Leone, where 41 percent of owl killings were simply for food. Unfortunately, non-selective eating also hits badly some very rare species like the Anjouan Scops Owl (*Otus capnodes*), Grande Comore Scops Owl (*Otus pauliani*), Mohéli Scops Owl (*Otus moheliensis*), and Rufous Fishing Owl (*Bubo ussheri*). In Africa, owls and their body parts are also used for traditional medicine, representing 6 percent of killings. There are tens of ways how the owls are or have been used as traditional medicine. In recent times, many of these practices have fallen away or substitutes have been found, but some of the traditional uses persist because people believe they work. Primarily, owls were killed for magic and witchcraft, explaining 28 percent of the reported cases. Six interviewed witch doctors admitted that they had used owls for bewitching and even killing people often combining owl parts with poisonous plants and a lot of magic. Now, some of them said that owls should not be used for killing, because the victim suffers too much when death can take four weeks or more. By using lion, leopard, or even snakes and crocodiles, death is instant. The haunted house story from Mozambique is an example of how extreme owl superstitions still affect many people's lives in Africa.


**Keywords:** rare owls eaten, traditional medicine, magic and witchcraft, haunted house

## 1. Introduction

There is a strong cultural link between humans and birds. Few birds or animals have gathered so many different and contradictory beliefs about them: Owls have been both feared and venerated, despised and admired, considered wise and foolish, and associated with witchcraft and medicine, the weather, and births and deaths—and have even found their way into *haute cuisine* [1].



IntechOpen

Folklore has it that owls are birds of ill omen and that deception is one of their favorite ploys. As a counterbalance, it must be said that the owl has been widely admired through the ages by deities, scholars, poets, and animal lovers in general [2].

For centuries, people have been using indigenous birds, not just for food and decorative purposes, but also to treat illnesses and to help them deal with difficult and often unexplained psychological and spiritual affairs. Birds have been used to treat physical and mental illnesses, strengthen relationships, encourage good luck, help promote dignity, cast spells, and strengthen individuals and give them protection against evil spells. Mankind has built up a whole range of superstitions about owls, a curious mixture of feelings in which the owl figures as a good or bad creature. Common to many societies is the belief that owls have superhuman powers [3].

There are few regions of the world where owls have had quite the impact that they have on African societies. Throughout the entire continent, there is a complex, sometimes contradictory, but more often corresponding, body of mythology and folklore centered on these birds [4]. For this chapter, I have collected mainly African examples of how owls are used as food, in traditional medicine, or as means of sorcery.

## 2. Methods

The author spent between 1977 and 2010, almost 22 years, in Africa being a resident in six different countries and a workwise visitor in additional 40 countries. All that time, owl beliefs were recorded to some extent, but more detailed interview studies were undertaken in 20 central, eastern, southern, and western African countries by using a pretested interview form in English and Portuguese in Angola and Mozambique (see [5]).

A local person, Bob Milingo Mvula, undertook the most sensitive interviews with the randomly selected contemporary witchdoctors in Malawi, and a female wizard was interviewed by the author in the Gambia. Handwritten notes were kept from each interview following the Annex 1 questionnaire. A total of six male sorcerers were interviewed in Malawi on the use of owls for bewitching people. The sensitivity of the interviews came from the fact that all respondents thought first that Mr. Mvula wanted to become a witchdoctor himself. For that reason, they were reluctant to reveal the secrets, especially the ingredients of their poisonous bewitching medicine. Some thought that Mr. Mvula was too young to become a witchdoctor—so in some interviews, he had his old father with him. If he admitted that he was collecting the information for a "white man's" book, then all respondents required money for these secrets, but we did not have too much money for this. As one Malawian lady said before refusing the interview: "I can't give information without some money. This time things are no longer like in the past when whites used to collect information for nothing" [5]. During the 22 years in Africa, the author collected all noted witchdoctor-related newspaper articles, out of which some anecdotes have been published before [6–14]. **Table 1** gives the details of how owls were used as food, in traditional medicine, or as means of sorcery based on 794 interviews and 333 recorded owl killings in Africa.

**Table 2** shows further details on owl species known to be eaten in Africa, also indicating the rarity status of the species. **Table 3** has a global list of some known owl-related traditional medicines. It is important to note that the effectiveness of any of the listed medicines from owl parts has *not* been scientifically proven, nor even studied, and their potency in all cases may be more than questionable. They are

| Region or country | Bad omens and evil spirits | Simply as food | Traditional medicine | For or against the witchcraft |
|---|---|---|---|---|
| East Africa (51) | 25 | 12 | 8 | 22 |
| Malawi (41) | 29 | 17 | 5 | 11 |
| South Africa (68) | 29 | 25 | 9 | 19 |
| West Africa (48) | 11 | 19 | 4 | 12 |
| Gambia (98) | 4 | 4 | 4 | 59 |
| Sierra Leone (27) | 15 | 41 | 4 | 11 |
| Total (333) | 17 | 16 | 6 | 28 |

*In this table, the number of killings is indicated in parentheses after the region or the country (countries with a lot of killings are presented separately from the region). East Africa = Botswana (6), Kenya (16), Lesotho (6), Mozambique (9), Tanzania (9), Uganda (5), Zambia (18), and Zimbabwe (15); Malawi (147); South Africa (181), and West Africa = Angola (4), Cameroon (6), Ghana (20), Guinea (8), Liberia (14), Namibia (4), Nigeria (96), and Senegal (5); the Gambia (188); Sierra Leone (37) (total number of the interviews in parentheses). Percentages above include only four reasons for killings, excluding many others like killing for decorations, feathers, fun, mistake, or due to the noise of owls.*

**Table 1.**
*How many owls are killed because they are bad omens or evil spirits or to be used as food, traditional medicine, or witchcraft in Africa as a percentage of all listed killings.*

| Owl species | Status | Country where eaten | References |
|---|---|---|---|
| Grande Comore Scops Owl *Otus pauliani* | Endangered | Grande Comore Island | [16] |
| Anjouan Scops Owl *Otus capnodes* | Critically Endangered | Anjouan Island in the Comoros | [17] |
| Mohéli Scops Owl *Otus moheliensis* | Critically Endangered | Mohéli Island in the Comoros | [16] |
| Pharaoh Eagle Owl *Bubo ascalaphus* | Least concern | Northern Africa | [15] |
| Spotted Eagle Owl *Bubo africanus* | Least concern | Sub-Saharan Africa | [15] |
| Fraser's Eagle Owl *Bubo poensis* | Least concern | W-Africa | [15] |
| Verreaux's Eagle Owl *Bubo lacteus* | Least concern | Sub-Saharan Africa | [15] |
| Rufous Fishing Owl *Bubo ussheri* | Vulnerable | Nigeria | [18, 19] |
| White-browed Owl *Athene superciliaris* | Least concern | Madagascar | [15] |
| Madagascar Owl *Asio madagascariensis* | Least concern | Madagascar | [15] |

**Table 2.**
*Eating and killing the rare owl species in Africa. Owls listed in taxonomic order [15].*

| Disease or problem | Owl parts needed | Medicine preparation | References |
|---|---|---|---|
| Alcohol abstention | Owl egg | A child will never be a drinker if eating an egg | [20] |
| Against epilepsy | Owl eggs | Soup made when moon waning | [20] |
| Against snakebite | Owl feet | Burn with herb Plumbago | [20] |
| Aphrodisiac | Owl meat | Meat must be eaten | [21] |
| Asthma | Body of an owl | Cures it since owls eat coffee beans | [20] |

| Disease or problem | Owl parts needed | Medicine preparation | References |
|---|---|---|---|
| Earache | Owlet's brain or liver | Mix with oil and inject into the ear | [20] |
| End fever | Right leg | Just to have a right leg | [21] |
| Eye complaints | Owl eggs or entire owl | Must be charred and powdered | [20] |
| Give dislike of wine | Owl egg | Eating one egg gives a lifelong aversion to wine | [20] |
| Gout | Owl body without feathers | Mummify in the oven, mash, mix with pig fat and apply to the affected site | [20] |
| Gray hair | Owl eggs | Use an egg to darken the hair | [20] |
| Hemorrhage | Entire Barn Owl | Boil in oil and add ewe-milk butter and honey | [20] |
| Hair-loss prevention | Owl eggs | A good cure for thinning hair | [20] |
| Hangover cure | Owl eggs | Cook eggs three days in wine | [20] |
| Hypnotizing | Owl's blood | To be used internally | [21] |
| Improve appetite | Owl meat | To be eaten as a delicacy | [22] |
| Infection of sinews | Long-eared or Eagle Owl's head | Take ashes with lily root and honeyed wine | [20] |
| Influenza | Magical owl hooting | Strain to hear cures worst symptoms | [20] |
| Night vision | Owl eyes | Eyes must be eaten | [23] |
| Rheumatism (1) | Owl feathers | Burn over charcoal | [20] |
| Rheumatism (2) | Owl meat | A gel made from owl meat | [21] |
| Seizures in children | Owl eyes | A broth made from owl eyes | [21] |
| Stop the child from crying or help them sleeping well | Owl feathers | Put feathers under the pillow. Works for children and adults as well | [20] |
| Whooping cough | The entire body of an owl | To be eaten as a soup | [20] |

**Table 3.**
*Some traditional owl medicines as known from Africa, Europe, India, and South America [20–23].*

presented in **Table 3** only as a curiosity and as an example, which may explain the thinking behind any unnecessary killing of owls.

## 3. Owls on the menu

Although the body of an owl is believed to have some magic power, not everyone in Africa finds eating them agreeable. South African Bantus say: "Asituiwa kuba silixelegu; umzimba waso uzele yinkwethu, into ke leyo sinuka ngathi sifile" (it is not eaten, for it is a slut; its body is full of scurf, which causes it to smell as if it was dead) [23]. But if you are starving, the taste of the food is not the most important issue as is later shown in the case of Sierra Leone.

East Africa is well known for its high wildlife awareness because of the importance of tourism. However, owls are still commonly killed for several reasons. Twelve percent of 51 recorded killings used the owl simply as food or to make condiments, or the killer just wanted to taste the owl meat (**Table 1**).

In South Africa, 34% of respondents knew people killing owls mainly for traditional medicine and eating reasons (**Table 1**). Some older people refused to answer this question, maybe indicating that in their youth, eating owls was much more common.

In the Gambia, more than half (59 percent) of respondents knew of people killing owls for or against witchcraft, but only 4 percent had witnessed somebody eating owl meat or even had done so themselves (**Table 1**).

Elsewhere in West Africa, most likely dreadful civil wars in Sierra Leone and Liberia explain that 19% of 75 reported owl killings involved "to be eaten" as the main reason, but owls were killed simply for food also in Senegal and Nigeria. In Sierra Leone alone, 41 percent of the killings were only for food (**Table 1**). Before eating the owl, people just catch them and burn the feathers, put salt, and roast the meat. Creoles of Louisiana, perhaps inheriting dishes from West Africa, used to eat Barred Owls (*Strix varia*)—but since no recipes have been handed down, we can only surmise whether the bird was roasted or put in a stew [1].

There are several statements about how very rare owls are still commonly eaten in Africa or the adjacent islands. It is said that the principal threat to the Anjouan Scops Owl (*Otus capnodes*) is the fact that it is still widely captured for food [17]. However, it is too rare to be worthy of special searches, but any owl encountered is usually taken, and hunting can be considered the main threat to its survival [15]. The same fate concerns the Grande Comore Scops Owl (*O. pauliani*) [16] and Mohéli Scops Owl (*Otus moheliensis*) (**Table 2** and **Figure 1**).

Nigeria got a lot of global publicity when in Ebu State in October 1997, a fisherman was seen eating a rare and vulnerable Rufous Fishing Owl (*Bubo ussheri*) [18, 19]. He admitted taking advantage of a soaked owl rendered unable to fly in the early morning.

**Figure 1.**
*Critically endangered Mohéli Scops Owl (Otus moheliensis) tied with a rope before being eaten. Photo: Courtesy of René-Marie Lafontaine.*

The owl was killed after being hit with the paddle. The fisherman claimed that the Rufous Fishing Owl was common in the area. Since that killing, he has turned out to be a converted protectionist of this globally threatened owl species in the area [24].

A comprehensive study of owls and humans was made in Malawi [5, 24, 25], and it seems that the owl meat is mainly used for bewitching and killing people, but not so much for eating as food. Every fourth of 147 respondents knew people who were believed to eat owls, and one out of ten had seen people eating owls. Males more often witnessed such happenings than females. Owl eaters existed both in cities, like Lilongwe, and in typical villages, but often, northern respondents felt that owls were eaten in southern and central regions only and naturally *vice versa*. Half of the interviewed people knew somebody who had killed or sacrificed an owl. Therefore, killing and sacrificing owls is much more common than eating them for food. Again, males were more aware of the killings than females [5].

## 4. Owls in traditional healing

There seems to be a worldwide consensus that owls can be used as medicine for healing some diseases, although the effectiveness of any owl part of medicine has not been proven or even studied. Many of the folk remedies survive and reappear throughout the world at different times. How we all happen to hold similar beliefs about parts of the owl's anatomy is partially a mystery. In some cases, the transfer of such beliefs is easy to trace; for instance, many of those held in medieval Europe originated in Greece and Rome, and those held in North America would have traveled to that continent with, say, the slaves of Africa and, later, with European immigrants [26].

The body parts of owls are used in both curative and preventive medicines (**Table 3**). In curative medicine, they are mixed with herbs and/or parts of other animals and given to patients to ingest, use as a lotion, or place on burning coal and inhale the smoke emitted. In preventive medicine, the preparations need not be in contact with the body. The owls or their parts may be buried at a chosen point around the home to keep away bad spirits and illness or be carried in pockets or bags as good luck charms [27].

Not so many details have been written on the use of owls in folk medicines, although some medicinal applications seem to exist for a variety of owl products. Different parts of the owl's anatomy have been and are highly prized as ingredients for indigenous systems of medicine [27].

In Germany, the first natural history and medical encyclopedia was published in 1491 [28] in which it was recorded that treatment for madness included the placing of owl ashes on the lunatic's eyes. This attempted cure was doubtless based on the principle that the owl's wise vision could, in this way, be infused into the madman's wildly distorted vision [29].

An interesting medical belief is that eating raw eggs of owls would cure a person of drunkenness (**Figure 2**). The eggs of an owl should be broken and put into the cups of a drunkard, or one longing for drinks; it will work in such a way that he will suddenly loathe his liquor and be displeased with drinking [30]. In Gironde, France, the same cure involved an omelet made using 5, 9, or 13 owl eggs. The belief presumably came into being because the owl is such a studious, solemn-looking bird that it was felt to epitomize sobriety and therefore to lay sobering eggs [29].

Greek writer Philostratus (ca. 170–250 AD) said that owl eggs made into soup as the moon wanes was the cure for the sickness, and when given to children, it would ensure lifetime sobriety and temperance. It was also presumed that one could restore

**Figure 2.**
*Owl eggs have been used to cure a person of drunkenness. Photo: Grayish eagle owl (Bubo cinerascens) eggs/ courtesy of Clive R. Barlow.*

clear thinking or reason after a night's boozing by consuming owls, especially their eggs. People would eat the eggs beforehand as a prophylactic against drunkenness (Mark Cocker, *in litt.*).

A variation of the owl-eggs-for-curing-drunkards theme saw the eggs administered repeatedly in glasses of wine (**Figure 3**). At first glance, there appears to be a basic flaw in this version of the treatment, but then again, perhaps the eggs made the wine taste so vile that even this method eventually worked [29].

One much-acclaimed cure for gout (in earlier times, it was believed to be only the classic boozer's complaint brought on by over-indulgence in alcohol) calls for the sufferer to take an owl, pull off its feathers, salt it well for a week, then place it into a pot and close it with a lid, and put it into an oven so that it may be made into a "mummy." It is then finally mashed, mixed with boar grease, and smeared liberally onto the affected place (**Table 3**).

There is also an old belief that owl soup will help to cure whooping cough. Again, the entire body of an owl is needed to make the soup. This is based on the idea that owls hoot and whoop so much without doing any harm that a broth made from their bodies should cure the disease [27, 29]. Today, due to animal protection, the genuine recipe may, in any event, be hard to come by, but the authentic version is said to beat any modern medicine!

There is a common Indian belief that the meat of an owl is a potent aphrodisiac [21]. On the other hand, in other places and at other times, this "tasty" morsel could cause loss of memory and, if you are particularly unfortunate, result in complete insanity. Also, in India, the eyes and flesh of owls cure some diseases. For instance, seizures in children could be treated with a broth made from owl eyes, and rheumatic pain is treated with a gel made from owl meat. A nomadic Kuravar tribe in India also stated that owl meat, particularly liver, eyes, and flesh, could cure the diseases associated with lungs and eye-related problems [29, 31].

**Figure 3.**
*Glasses of wine and owl eggs could cure drunkenness—Especially owl wine. Photo: Courtesy of Rudolf Schaaf.*

Clouston [31] reported that owls' eggs and the blood of their nestlings preserve the hair and make it curly. The Shoshone Indians in the United States believed that dandruff could be cured by simply putting your head inside the burrow of the burrowing owl (*Athene cunicularia*) and shaking your head.

The ash obtained by burning owls' feet together with the herb plumbago is considered a nostrum against snakebite. This is not quite as silly as it sounds because owls kill with their feet and their legs; feathering also gives some protection against prey bites, so people might be justified in thinking it would be just as effective against snakes. Along these lines, there was another bizarre medication in Germany that helped you to avoid being bitten by a mad dog and contracting rabies. You must place the heart and right foot of an owl under your left armpit. Luckily, people in Germany have stopped placing parts of owls under their armpits, thanks to modern medicine against rabies [29, 31].

But in Africa, it is still a general belief that eating the eyeball of an owl would give a person night vision due to the owl's ability to see in the dark. It is interesting that also in northern India, it is believed that if one eats the eyes of an owl, they would be able to see in the dark [21]. In Zambia, for instance, the dried eyes of owls are mixed with other traditional medicines and used by hunters so that they can see clearly and aim at targets during night hunting. According to North African belief, it is essential to know which eye to use, for one eye of an owl sleeps, but the other is permanently wakeful. To tell which one is which, the eyes must be put into a bowl of water; the sleepy eye will sink, while the other will float [1]. In ancient Europe, it was not necessary to eat the eyes, but only applying an owl's tears to a person's eyes enabled that person to see in the dark [32].

In Algeria, owls were used to cure blindness and headache [33]. The advice was to go to an owl's nest and blind all the young ones, and then line the nest with cotton wool. When the female owl returns, she will bring medicines to cure her young, and some of that medicine will fall upon the cotton wool. That can be removed later and used on the human eyes. If the bandage is soaked in a special lotion (fidili + runhu) and tied around the head, it will act through the eyes and cure headaches. Thus, some Tripoli women tie this kind of bandage continuously to their hair on the right side and never suffer from headaches [33].

It is a popular belief in Morocco that the Barn Owl (*Tyto alba*), and often also Little Owl (*Athene noctua*), is thought to cure skin diseases; the birds are killed, cooked, and eaten. It is also believed that "powdered owl," applied as an ointment, is the ideal cure for eye complaints. The remedy for earache is by injecting into the ear an owlet's brain or liver, mixed with oil, or by applying the mixture to the parotid gland. Owl feathers are also of use to man—when burnt over charcoal, they are a good cure for rheumatism. In India, rheumatism is also treated with a gel made from owl meat (**Table 3**).

Commonly, owl's egg soup was reckoned to be effective against epilepsy, the only snag being that it had to be prepared when the moon was on the wane. Folk medicine advocates strain their ears to hear the magical hooting of an owl, guaranteed to banish the severest symptoms of influenza. Again, like in India, seizures in children could be treated with a broth made from owl eyes (**Table 3**).

In South Africa, owls are still used for healing purposes, and it is estimated that 70 percent of the black population makes use of the services of traditional healers [34]. Traditional medicine or "muthi" is a billion Rand business [35]. The apartheid era placed restrictions and censures on this industry, which forced most traditional healers and sangomas to operate in secrecy (Suppression of Witchcraft Act of 1957). The post-apartheid era has allowed traditional healers and the use of traditional medicine to venture into the limelight. Traditional "muthi" medicine markets are now found in all major cities and throughout rural villages within South Africa. The decriminalization of the traditional medicine markets and sangomas in post-apartheid Africa has resulted in calls from traditional healers to be afforded greater recognition within the medical fraternity. These calls extend from the ability to advertise themselves as doctors to the issuing of valid doctors' notes for illness-related absenteeism within the workplace [35].

As in many cultures globally, owls have featured prominently in South African folklore and mythology. The use of owls and owl parts, although acknowledged, has to date not been qualified or quantified within the traditional healers' "muthi" chest. Through site visits to "muthi" markets in South Africa and the partition of traditional healers in a simple questionnaire, the uses associated with owls and their body parts have emerged. In addition, insights have been gained into the collection and identification of owls by these healers. The so-called "muthi" owls in South Africa are distinct as they are used for both traditional cures relating to headaches and insomnia as well as for spiritual curses and "witchcraft" [35].

Mrs. Zodwa Khumalo is one of those healers in Durban. According to her, there are a variety of beliefs surrounding owls. Alive, they may be a harbinger of bad news or evil spirits, but once dead, they can be used to help people with little energy who sleep during the day. She tells people: "You must wake up in the day. You are a person, not an owl." And then, they must take some medicine made with the owl [34].

With the influx of both legal and illegal migrants in South Africa through undefined borders, the reliance on traditional healers and their remedies will likely continue to rise [35]. Also, in Zimbabwe, the traditional medical business is booming, and this has had serious implications for raptors and other birds used in medicine [36].

In Malawi, south-eastern Africa, where a comprehensive study of owl beliefs, legends, and myths was done [5, 24], owl-based medicines were mainly used for bewitching and killing people, but not so much for healing many diseases.

The idea that owls can be used as medicine for healing some diseases is based on the traditional precepts of sympathetic medicine, whereby eating an animal or parts of it enables the patient to not only benefit from the meat itself but also absorb the physical and moral characteristics of that creature's sharp night vision, very good hearing, and the like, as mentioned above.

Therefore, it is unlikely that traditional healing will end soon. It has even been increasing at least in Africa, which has alarmed many people interested in birds from a scientific or recreational perspective [37, 38]. However, more owls are likely killed on roads by motor vehicles, or lost through habitat destruction, than are sold in medicinal or "juju" markets. But by ensuring a supply of owls for traditional healers, environmentalists could also conserve certain vulnerable species.

In 2000, I raised a not-so-well-defined suggestion that common, easily managed species of owls could be bred in captivity to satisfy some of the demands for body parts of owls from practitioners of traditional medicine. I assumed that this would provide legitimate, readily available material, which would reduce some of the pressure on much rarer species in the wild. My biologist wife, Anita, found this suggestion repugnant and so did many other readers of my paper [39].

After the wife of Bernard Sayers visited the commercial farming of non-domestic animals (crocodiles and tigers) in Thailand, Bernard commented [40] on my wild suggestion and found it a little more acceptable should it reduce the threat faced by so many rare owl populations. The owl keepers could produce enough barn owls for medicinal purposes and thus avoid the unnecessary killing of African Marsh Owls (*Asio capensis*), Madagascar Red Owl (*Tyto soumagnei*), or African Grass Owl (*Tyto capensis*) and other rare species because a medicine man does not know the owl species.

It has been noted in Africa that tourists are flocking the stalls of medicinal street markets and finding how African people use indigenous materials fascinating. If we accept the fact that crocodiles and rhinos are farmed, why not rare birds and owls?

Bernard Sayers [40] concluded his writing on my suggestion: "So shall we see Barn Owls or other species of owls farmed commercially to supply the international trade in traditional medicine. I do not pretend to know the answer, and, in many respects, I hope it does not happen, but if through well-run commercial farms the temptation to take much rarer species from the wild is reduced then I guess that it must be the lesser of the evils. Should such an operation arise the critical issues are to ensure that farms are well run, the birds well cared for and the conservation benefits maximized." Then he concluded: "I appreciate that commercially breeding non-domestic animals for slaughter is an extremely emotive and highly sensitive topic and it would be interesting to learn the views of other members (of The International Owl Society)." Due to very strong points of my wife on encouraging questionable medical treatments, I had already decided not to say a word about this suggestion any longer.

However, recently, Haw [35] has given almost similar suggestions to be explored. Wildlife rehabilitation centers and zoological gardens routinely receive owls that require euthanasia, which could contribute to "muthi" markets and this way reduce the exploitation associated with the wild harvesting of owl species by traditional healers.

## 5. Owls and witchcraft

Many of the ideas revolve around two key beliefs. Owls are perceived as harbingers of evil, and usually, they are viewed as messengers who announce forthcoming illness or even death to the observer or the observer's family [1]. The other central belief is that owls are integrally involved in witchcraft. Should an owl appear to a person steeped in traditional lore, then the individual often makes one of two assumptions: either the owl is seen as a nocturnal form assumed temporarily by a witch during his or her nefarious practices, or the owl, while retaining its natural form, has somehow been coerced into service and is itself abroad on the witch's business. What is of significance is the fact that these negative ideas are widespread, have great potency, and are deeply rooted in parts of indigenous societies and even in contemporary Africa, the Middle and the Far East, and parts of South America [26].

**Table 1** shows that owls are widely killed and collected in Africa to supply magico-medicinal practices often connected with witchcraft. The witch doctors are required during the so-called "blood brotherhood" initiation, to mingle their blood with that of an animal. And whereas the blood bond is likely to be established with one of the more ferocious species, such as a lion, leopard, snake, or crocodile, night birds such as the owl are an acceptable alternative. Witches proper are renowned, of course, for preparing a variety of brews and potions, which form the crucial ingredients of their spells and the very basis of their magic [1].

Six contemporary medicine men (proven to be real witchdoctors) were interviewed in Malawi in 1997.

1. Dr. E. Kamwendo, as the name is written at a place in Lilongwe Market, where he operates, was interviewed on January 23, 1997. At that time, he had been a medicine man for the past 30 years and had an interesting recipe for preparing the bewitching medicine. Kazizi (an owl; also a Nyanja name for the Spotted Eagle Owl [*Bubo africanus*]) is killed—this can be done by a vehicle or by any other means. The fat is taken from the owl and mixed with "*mwavi*," which is a strong tree bark poison from Forest Ordeal Tree (*Erythrophleum suaveolens*).

In old times, "*mwavi*" liquid was used to define who is a witch or a wizard. People drank the liquid, and those with weak blood died and those with strong blood only vomited. People believed that those who died were witches. In Dr. Kamwendo's bewitching medicine, some leftover "*nsima*" (maize porridge) and gravel from a fresh grave are added; the last mentioned has to be collected secretly. The last ingredient can be only a small stone, and everything is then pounded.

At night, between 9 and 10 pm, this preparation is taken to the roof of the intended victim's house. The witchdoctor stands on the roof at one corner of the house and throws some of the powder into the middle of the roof, then into the different corners of the house, and lastly again into the middle of the roof. Then early in the morning, an owl will be on the veranda or at the front of the house. The owl is captured and taken to the graveyard at night and told that this is his home!

Dr. Kamwendo was reluctant to reveal further his secrets about how the owl was then used for killing from the graveyard but asked us first to prepare an owl by using his method. Anyhow, he concluded that the power of his owl was the same as that of a poisonous snake. Later, Mr. Mvula visited Dr. Kamwendo again to ask what the owl that was created by the witchdoctor ate. He said that it ate what the normal owls ate. He added that even the lion that has been created by people (man) ate what the lions that God had created ate.

2. On January 25, 1997, Mr. Mvula interviewed the witchdoctor known by the name Balowe Shaba. Balowe means letting people bewitch him or her. By naming himself Balowe, he meant that even if people could bewitch him, nothing was going to happen to him. Mr. Shaba told that the owl parts are used in several ways; the parts he mentioned were the head and heart. A head or heart of an owl is taken and a place is chosen at the yard of a house, where the pounded owl part is put under the soil. It must be well prepared for planting a fruit tree or any other plant, which he will choose. The plant or seed of a fruit tree is planted so that the growing plant will pass through the middle of the pounded owl's head or heart. After this plant or fruit tree has grown enough, its roots can be used for bewitching. Before using it, the root is mixed with some other roots or herbs, which he did not reveal. When the witch-doctor wants his medicine to go and bewitch someone, he will talk to the root and some other roots, which will turn into an owl. The major component of the magic is called "*mphiyi*." These are small pieces of a branch of a tree or a root of a tree that are collected, and the outside part is removed and then cut into pieces of more or less one inch. When sending the bewitching medicine, the witchdoctor speaks to the "*mphiyi*" by saying: "I am sending you to such and such a person, to do such and such a job, because I want him or her to die." Immediately after the witchdoctor finishes talking, the "*mphiyi*" turns into an owl, which then flies to the person to whom it has been sent. The owl lands at the house of that person and hoots. After a few days, the person or one family member starts suffering due to some illness, which eventually kills that person some days or weeks later. Mr. Shaba confirmed that as long as the owl is kept in the form of "*mphiyi*," no feeding is needed. He also added that no part of an owl is used as medicine for healing any diseases. The owl is collected dead or alive; a road kill can also be used. Mr. Shaba felt that the strength of the owl-based medicine only differed slightly from potions prepared by lions, leopards, or reptiles.

3. On January 26, 1997, Mr. Mvula visited Mr. Msamale Wjiko, who was first very suspicious of what Mr. Mvula wanted from him. Later, he opened up and said that the owls were kept in the form of "*mphiyi*" and that "*Chizimba*" was the major component for the bewitching medicine to work. Chizimba means a substance made from a special part of a wild or domestic animal or a living creature. In this case, owl parts included are the heart, brain, legs, and beak, which are mixed with "*mono*," the black one, and "*kachebele*" from the river. "*Mono*" literally means a basket to catch a

fish—but it is assumed that the black one is a fish. "*Kachebele*" is a local name for the common Water Hyacinth (*Pontederia crassipes*), which is considered toxic to humans. These elements are put together and ground. He also uses "*mwavi*," which is a strong tree bark poison from the Forest Ordeal Tree. That tree is found at Choma in Mkubazi Mountain and Vizala Rubber. It is the most powerful tree, so when taking it, the medicine man must not sleep with a woman for one month. "*Mwavi*" is added to the above-mentioned items, and all elements are burned together. Some other medicines are still added, but "*mwavi*" alone can kill a person. If the tree bark is taken, burnt, and ground and this powder is added to some food or beer, the eater or drinker will die. Another way of using the bark is just by drying it, after crashing it, then grinding it, or just soaking the bark in the water. This stuff and some "*mphiyi*" are put on a piece of a broken pot of clay, which is called "*dengere.*" This piece of clay with medicine on it is put on fire, and the snout is taken from the nose. "*Chimphonogho*" (snout) has two functions: One is for rubbing the medicine, and the second is for sending the medicine to where the owner wants. If the snout that is used for sending is taken from the right side of the nose, then the owl will be a male, and if taken from the left side of the nose, the snout will form into a female. If snout from both sides of the nose is mixed, then two owls will be made, and they will go to the place where the owner wants to bewitch the desired person. According to Mr. Msamale Wjiko, the female owl is more powerful than the male owl, showing that he is well aware of the sexual dimorphism in owls [20].

The "*mphiyi*" is rubbed with the oil of black "*mono,*" and the stuff is kept in a house until the day of use, when it is taken to a place where the rubbish (e.g., ashes) is thrown. Then, one has to wait and see if the medicine will turn into one owl or two owls. To improve the results, a song is sung, "*kazizi-kula-kula*" (owl-grow-grow). And the owl grows, and it becomes big, looking like a cat or a bat.

Mr. Msamale Wjiko also confirmed that owls were not used for healing medicines, and according to him, animals like leopards and lions, and even snakes are more powerful than owls. With the animals and snakes, the person who is a target is killed at once, whereas when using the owls, the victim suffers between 1 and 4 weeks.

4. On January 27, 1997, Mr. Chimutoto Mdhluli was interviewed. He was feared as a witch, and even Mr. Mvula went to see him with his old father. First, he only saw the nephew of Mr. Mdhluli, who is also a young medicine man. He was very suspicious toward Mr. Mvula and did not want him to write down anything. Later, the interview materialized with Mr. Mdhluli himself. He stated that an owl was not used as medicine for healing but only for bad things like the bewitching of people. He uses the roots and stem of a tree called "*Muchemani,*" a Latin or English name that is not known to us. Another new term in his magic was "*kutembelera*" (a curse word in Chichewa), which is used when talking to the roots or stem of the "*Muchemani*" tree. This tree is used in many ways: for the growth and prosperity of a business, for women to be loved by their husbands, or for males to be loved by women. People say that a tree has the power to attract something (e.g., women, if the user is a man).

After "*kutembelera,*" the owl in the bush becomes wild and goes to where it has been sent by the medicine man, the owner of the "*mphiyi.*" According to Mdhluli, lions, snakes, and leopards are more powerful than owls. The owl tells that something bad is going to happen in the village. The medicine man, like Mr. Mdhluli, can protect the house to avoid owls and any other form of witching to happen at the house of a person who does not want to bewitch. In the case of Mr. Mdhluli obviously, no owls are created or taken from nature, but only through "*kutembelera*" (cursing) by using the "*Muchemani*" tree.

5. The fifth medicine man interviewed was Mr. Msamale. His name means "*watch out,*" meaning that witchcraft people should be careful with him. He was the only interviewed person who told us of the use of owls for something other than killing people. His owl is used to save somebody who has committed a serious crime or to release someone already in prison. His medicine preparation is as follows: He goes into the valleys or gullies looking for a nest of an owl, which has young ones. The leg of a young owl is tied to a branch of a tree or to a tree itself. When the female or male owl comes back to the nest, it sees that a strand of bark from the rope tree tethers its young. The strand that is used is made from the bark of a "*mtondo*" tree, which is the sunbird or wild mango tree (*Cordyla africana*).

The parent owl then goes to look for a certain tree. The root is collected, and when it returns to the nest with that root, the strand tying the young owl cuts itself. Then the medicine man comes to collect the root that was brought by the parent owl. This same root is used for freeing the person from jail or to stop the jury from finding the accused criminal guilty of any crime. And the accused goes free.

6. Mr. Mvula tried to interview one more sorcerer called Mr. Moyo in Lonje but found such poverty in his house that he gave up by concluding: "I hate poverty." The sorcerer's house had three almost naked children and two wives. Boiled cassava was the only food, but when that was offered to Mr. Mvula, he found it below his standards. Mr. Moyo was suspicious and stated that he knew nothing about how the owls were used for bewitching people. Witchdoctors in the area were afraid to talk because some traditional healers had accused them of sorcery.

As seen above, all witchdoctors had slightly different ways of using owls, but in almost all recipes, the owl was similarly recreated through a complicated ceremony combining poisonous plants and a lot of magic. The owls are collected dead or alive, often they are road kills, but sometimes, owls are also especially killed for the bewitching medicine. If seen catching an owl, the medicine man tells suspicious people that he is going to eat the owl. The species of owls are not mentioned and do not make any difference in witchcraft. Indirectly, however, two species became known and are often used, Barn Owl and Spotted Eagle Owl. When the owl is in this "*mphiyi*" form, as it is called locally in Malawi, it will not need feeding according to some medicine men. Others say that owls are owls and eat what they normally eat.

The owl is used in several ways, but most often the head or heart is put into some place in the garden, which the medicine man has chosen. A seedling or seed of a fruit tree is planted into the pounded head or heart of an owl. When this fruit tree grows, its roots are used for bewitching. The root is mixed with some other roots or herbs, which are often poisonous. The medicine man or the owner of that fruit tree will talk to the mixed root and herb medicine, and the owl will be born again from the roots of the fruit tree. At the time, the medicine man sings a Nyanja song: "*Kazizi-kula, kazizi-kula-kula*" (grow owl, grow-grow). And the owl grows and becomes a big creature like a cat or a bat. The owl then flies to the person to whom it has been sent and lands on the roof of the house of that person. Then it hoots, and after a few days, the person or the family members start suffering from an unknown illness. Some days or weeks later, the suffering person dies. Almost hilariously, some interviewed medicine men concluded that owls should not be used for killing, as the victim suffers too much if dying takes 4 weeks or more. By using lion, leopard, or even snakes and crocodiles, death is instant. Again, healers disagreed with this.

During the interviews, Bob Milingo Mvula collected some plant remains that the witchdoctors had been using. Among those, we identified castor seeds (*Ricinus communis*). The plant is a shrub or small tree, and its seed (castor seed) contains ricin, one of the highly toxic substances in the world. One to three seeds can be fatal to a human being. The deadly ricin can be used in aerosol or pellet form as a bioweapon. Heat disarms the toxin. The plant is originally from Africa but is nowadays cultivated and naturalized throughout the world [41].

After the Malawi interviews, Mark Cocker studied traditional medicine markets and interviewed two traditional sorcerers in Cameroon [42]. Although there are no developed markets for dead birds or their parts, the few skins that are being traded by the ubiquitous herbalists are almost all owls. In Cameroon, the owl heart is identified as especially prized for its magico-medicinal power, although both sorcerers in that country confirmed that all parts of the owl were an effective medicine.

One explained that patients came to them because of a curse by another sorcerer whose intention was to make them ill. The aggressor takes owl parts and prepares a spell. To counteract this type of practice and work on the same principle as an inoculation, the traditional healer also takes owl parts, or a whole bird, and cooks it. The charred powder is then added to other herbal ingredients and given to the patient to eat. Any witchcraft in his or her body will be driven out.

Another sorcerer reported that owl parts were used in conjunction with other herbal ingredients to construct a form of talisman worn by the person in an amulet or similar container. This both served as a protection against witchcraft and could also be effective against general bad luck. Should an owl fly into the house of an individual wearing such a charm, the bird would instantly fall dead on contact with the "medicine" contained therein.

In 2002, a female diviner from Sierra Leone was interviewed. She was explaining how you get the owl to do even worse things for you, like killing a baby or so. First, you catch the owl and keep it secretly in the house. You give the owl palm oil, snuff, bené, and pepper (you must force-feed the owl). Then you explain everything to the owl, and after that, you strip yourself naked and go on the floor and start moving round and round. Then, the owl will understand what evil you want to do. After that, you release the owl. In the end, she added that owls were not bad, but if you understand the owl, you can make it do a bad job for you. When I asked if she is personally afraid of owls, she said: "Me, I can't be afraid of owls – if I want to enter an owl, I can do it and come back as a human after my evil mission as an owl."

Also, in India, owls and their body parts are primarily used for black magic [21]. There are probably at least 50 active wild bird-selling points/localities in India where any bird trader is likely to be able to procure owls, although the birds are rarely on open display. Owls are used to control a person or a lover with "*vashikaran*" (hypnotism). Both an owl's ear tufts and castor seeds are mixed with milk and then dried, powdered, and served in paan (betel leaf). The person who is served it, or has it sprayed on their head, will be hypnotized. Another method is using the heart of an owl, which is grounded or mixed with *gorochan* (a product from a cow's bile duct) in equal parts to form a paste. This paste is blessed seven times with a prescribed mantra, then used as an eyeliner, and the person targeted becomes hypnotized.

The live owls are used for countering evil eyes. If a person is having a spate of evil luck, then it is obvious that a jealous friend has cast an evil eye on him or her. So, what can be done? He or she orders an owl. When the owl is received, the person instructs the shopkeeper to kill the owl and then skin it. The shopkeeper is so eager to get his reward that he almost skins the owl alive. Then, the skin is burnt, and the flesh is buried, and the person goes home comforted by the fact that he or she has been liberated from the evil omen. Generally, Spotted Little Owls (*Athene brama*) are used for this purpose, and Abrar Ahmed recorded up to 20 owls for sale during a single visit to the major bird markets [21].

## 6. The "haunted" house story

I spent almost 6 years with the United Nations Food and Agriculture Organization in Mozambique, first before the war (1979–1981) and later during the dreadful civil war (1989–1993). Due to the demanding situation in the country, there was hardly any time to think about owls or owl beliefs when trying to find the fastest and best ways to feed hungry people. However, Mozambique gave us some owl experiences that prove that extreme superstitions exist in that country as well.

During work travel to Nampula, I saw a young Southern White-faced Owl (*Ptilopsis granti*) that had to be saved from the street market. That owl spent exceptionally long time with our family, often also traveling with us as we could not leave it alone in the house. So, it became an accustomed air traveler in an African basket made for it.

In the hotels, it was often free in the room and liked to sit at any higher point of the room (**Figure 4**). The only problem was that the cleaners panicked if they found an owl in the room, so during the cleaning, we had to hide it to avoid such occasions.

Being a VIP traveler, I was not normally forced to pass the security checks at the airport when entering the plane directly from the VIP lounge. Once, however, I went through the normal line at Maputo airport on our way to Swaziland (renamed Eswatini now), which was also a part of my FAO representation. The owl was passed in its basket through the security X-ray. The airport officer only saw the white bones

**Figure 4.**
*Southern White-faced Owl* (Ptilopsis granti) *on the window frame in Mozambique. This small, playful owl living on the balcony made the entire house haunted for the local people. Photo: Heimo Mikkola.*

of the owl on his monitor. Soon, this worried and blank-faced Black officer came to ask: "Are you a Diplomat?" I said "Yes," and the officer hushed us to go quickly to the plane. I still wonder what the security officer thought this "monster" to be or if he was able to recognize that it was a live owl.

On one of my family's regular Swaziland flights, my wife, Anita, realized that the owl had left the basket and was sitting on top of it. If anybody else would have seen the live owl, it could have caused a real panic inside the plane.

But this lovely little owl, the most talkative of any of the family's rehabilitated owl species thus far, caused a huge disaster after we safely returned it to nature in a wildlife park in Swaziland when we left Mozambique for a new duty station in Malawi. My successor took the same house in Maputo, where the owl had been kept in a large backhouse balcony where it was able to fly freely.

Dreadfully, this successor soon got ill after moving to Maputo and did not survive a tumor in his head. He was married to a local lady, who started to say that the house they had taken after our family was haunted because we had kept the owl in it. And that this was why her husband died. Luckily, we were safely out of the country, but this left an awfully bad feeling for the entire family. The lady refused to enter the house after the death of her husband, and other people were forced to move their furniture out from that allegedly haunted house.

We have never been to Mozambique after this but heard that the haunted house story came up again when the next successor took the house over and was seriously attacked at the gate of the house. During our time, we never had any housebreaking or larger robberies, but this could be because we had two dogs and one owl living with us. Weinstein also concluded that owls may be the cause of houses being described as "haunted" [1].

## 7. Concluding remarks

There are a few regions of the world where owls have had quite the impact that they have on African societies. By understanding the patterns of owl killings and use for witchcraft in old times and even today, we might be able to understand better our cultural past in this modern world. For witchcraft purposes, owls are collected dead or alive; often, they are road kills, but sometimes, owls are also especially killed for the bewitching medicine. Witchdoctors do not aim for any particular species and may not cause so much damage to the rarest species. More dangerous are people who catch the owls to be eaten or used for traditional medicine. The killing of the African Grass Owl, African Marsh Owl, Anjouan Scops Owl, Grande Comore Scops Owl, Madagascar Red Owl, Mohéli Scops Owl, and Rufous Fishing Owl, to name a few rarest species, can be devastating because the hungry person or medicine man does not know the owl species. The haunted house story from Mozambique is a sad example of how strong superstitions still govern many people's lives.

There is no scientific evidence to support any of the superstitions or traditional medicine's effectiveness to continue using, killing, or being afraid of owls. So, one would like to argue that although these traditions are very deep and difficult to change, education or sensitization programs can contribute to visualizing these species from a different perspective and so be able to protect them [43, 44].

There has been a remarkable project in Zambia, where the Ornithological Society organized the "Owls Want Lowing Survey" (OWLS) involving schoolchildren (7th–8th graders) from various tribes. Children asked their village elders about old stories and folklore on owls but had their educated views to complement the picture. The results were published as a book in 1999 [45]. Similar projects could be undertaken in other African countries to better understand and protect the owls.

# A. Appendix 1
## Questionnaire for the witchdoctor interviews by using the local languages, like Chichewa, Khonde, Nyanja, Tumbuka, and Yao.

### A.1 Personal Data

Name............
Age............
Female............
Male............
Education level............
How long has been a witch doctor............
In which Village/ City............
Origin............
How long present in the above village/city............

### A.2 Witchcraft-related Questions

Do you know any owl species?.........None.........Yes, how many.........
How the witch doctor gets his owls?.........
Dead or alive?.........
How the owl is used?
Which body parts are used to make medicine?.........................
What medicine..............................................................
Against what disease? .....................................................
Special ingredients of the poisonous bewitching medicine? .........
Plant species used? ..........................................................
How the witch doctor compares the use of owls with leopards......... lions............
snakes.......... crocodiles ......... other animals .........
Your comments on the interview?...........................................
Any other remarks? ..........................................................
Thank you for your participation and assistance in this study!

## Author details

Heimo Mikkola
University of Eastern Finland, Finland

*Address all correspondence to: heimomikkola@yahoo.co.uk

## References

[1] Weinstein K. The Owl in Art, Myth and Legend. New York: Crescent Books; 1989. p. 144

[2] Cenzato E, Santopietro F. Owls–Art, Legend, History. Milan: Arnoldo Mondadori Editore S.p.A; 1991. p. 112

[3] Redford KH, Godshalk R, Kiran A. Wild animal species in community forestry in the tropics. FAO Community Forestry Note. 1995;**13**:1-96

[4] Cocker M, Mikkola H. Owls and traditional culture in Africa. Tyto. 2000;**V**:174-186

[5] Mikkola H. General Public Owl Knowledge in Malawi. The Society of Malawi Historical and Scientific Journal. 1997A;**50**(1):13-35

[6] Mikkola H. Discourse. In: An Environmental Evening of the Wildlife Conservation Association. Maputo: Clube de Desportos da Costa do Sol; 1991

[7] Mikkola H. Los buhos en las supersticiones africanas. Quercus. 1995;**111**:25

[8] Mikkola H. Owls in African superstitions. Owls Magazine. 1995;**1**(**4**):2-3

[9] Mikkola H. Eulenaberglaube aus Afrika und Skandinavien. Kauzbrief. 1995;**4**(**7**):13-14

[10] Mikkola H. Africans destroy owl study site. Owls Magazine. 1996;**3**(**3**):19

[11] Mikkola H. The use of owl eggs in gambling in India. Tyto. 2002;**VI**(**IV**):171

[12] Mikkola H. Death of the owl. Tyto. 2002;**VII**(**I**):3

[13] Mikkola H. Weird Owls Rampage Kujube Village. Tyto. 2003;**VIII**(**I**):24-25

[14] Mikkola H. Oil of "Owlay" (sic). Tyto. 2004;**VIII**(4):12

[15] König C, Weick F, Becking J-H. Owls of the World. London: Christopher Helm; 2008. p. 528

[16] Louette M, Stevens J. Conserving the endemic birds on the Comoro Islands, 1: General considerations on survival prospect. Bird Conservation International. 1992;**2**:61-80

[17] Safford RJ. Rediscovery, taxonomy and conservation of the Anjouan scops owl *Otus capnodes* (Gurney 1889). Bird Conservation International. 1993;**3**:57-74

[18] Hall P. Pathos in Nigeria. Tyto. 1996;**1**(**2**):53

[19] Barlow RC. In search of the rufous fishing owl and other things. Tyto. 1998;**3**(**5**):138-143

[20] Mikkola H. Owls of the World Enhanced e-book. London: Bloomsbury/ Christopher Helm; 2014. p. 528

[21] Ahmed A. Imperiled Custodians of the Night: A Study on Illegal Trade, Trapping and Utilization of Owls in India. New Delhi: TRAFFIC India/ WWF-India; 2010. p. 78

[22] Wetmore A. Observations on the birds of Argentina, Paraguay, Uruguay and Chile. Bulletin of the US National Museum. 1926;**133**:1-448

[23] Godfrey R. Bird-Lore of the Eastern Cape Province. Johannesburg: Witwatersrand University Press; 1941

[24] Mikkola H. Owl Knowledge and beliefs in Africa. Tyto. 2021;**26**:9-35

[25] Mikkola H. Comparative study on General public owl Knowledge in Malawi and in eastern and southern Africa. Nyala. 1997;**20**:25-35

[26] Mikkola H. Búhos y humanos. Capitulo 5: 28–35. In: Chaparro Herrera S, Enríquez PL, Lopez-Salazar A, editors. Búhos de Colombia Guia Illustrada y Sonora. Bogotá: Grupo de Especialistas en Búhos Neotropicales. Puntoaparte Editorial; 2020

[27] Mikkola H. Owls in traditional healing. Tyto. 1999;**4**(**3**):68-78

[28] Meydenbach J. Hortus Sanitatis. Mainz, Germany; 1491

[29] Morris D. Owl. London: Reaktion Books; 2009. p. 216

[30] Swan J. Speculum Mundi. Cambridge: Roger Daniel; 1643

[31] Clouston WA. Notes on the folk-Lore of the raven and the owl. In: Saxby JME, editor. Birds of Omen in Shetland. London: Privately Printed; 1893. pp. 17-32

[32] Goddaeus C. Laus Ululae (the praise of Owls). Amsterdam; 1640

[33] Tremearne AJN. Bori beliefs and ceremonies. Journal of the Royal Anthropological Institute of Great Britain and Ireland. 1915;**45**:23-68

[34] Derwent S, Mander M. Twitchers bewitched: The use of birds in traditional healing. Africa–Birds & Birding. 1997;**2**(**1**):22-25

[35] Haw J. From cures to curses – Owls and their place in traditional healing in southern Africa. In: World Owl Conference. Portugal: University of Évora; 2017. p. 40

[36] Msimanga A. The role of birds in the culture of the Ndebele people of Zimbabwe. Ostrich. 2000;**71**(**1&2**):22-24

[37] Williams VL, Cunningham AB, Kemp AC, Bruyns RK. Risk to birds traded for African traditional medicine: Quantitative assessment. PLoS One. 2014;**9**(**8**):e105397

[38] Williams VL, Whiting MJ. A picture of health? Animal use and the faraday traditional medicine market, South Africa. Journal of Ethnopharmacology. 2015;**179**:265-273

[39] Mikkola H. Eulen in der Volksheilkunde. Facetten aus der Alten und der Neuen Welt: Die Eule in der sympathetischen Medizin und als Mittel der Hexerei. Kauzbrief. 2000;**13**:32-38

[40] Sayers B. The commercial farming of non-domestic animals. Guest Editorial in Tyto. 2000;**5**(**3**):97-99

[41] Zuchowski W. Tropical Plants of Costa Rica. A Zona Tropical Publications: Cornell University Press; 2007. p. 529

[42] Cocker M, Mikkola H. Magic, myth and misunderstanding: Cultural responses to owls in Africa and their implications for conservation. Bulletin of the African Bird Club. 2001;**8**(**1**):30-35

[43] Enriquez PL, Mikkola H. Comparative Study of General Public Owl Knowledge in Costa Rica, Central America and Malawi, Africa. In: Duncan JR, Johnson DH, Nicholls TH, editors. Biology and Conservation of Owls of the Northern Hemisphere: Winnipeg. St. Paul, MN: US Department of Agriculture and Forest Service; 1997. pp. 160-166

[44] Enriquez PL, Mikkola H. Creencias populares sobre los búhos en Centroamérica y Africa: Un estudio comparativo (Abstract/Resumen: Popular credence on the owls in Central America and Africa: A comparative study). Quercus. 1998;**150**:22-25

[45] Ashley N, Sichilongo M, editors. Owls Want Loving. Lusaka: The Zambian Ornithological Society; 1999. p. 48